Green Energy and Technology

For further volumes:
http://www.springer.com/series/8059

David Wood

Small Wind Turbines

Analysis, Design, and Application

Springer

Dr. David Wood
Department of Mechanical and Manufacturing Engineering
University of Calgary
University Dr NW 2500 Calgary, AB
T2N 1N4
Canada
e-mail: dhwood@ucalgary.ca

Additional material to this book can be downloaded from http://extra.springer.com

ISSN 1865-3529
e-ISSN 1865-3537
ISBN 978-1-4471-2706-2
ISBN 978-1-84996-175-2 (eBook)
DOI 10.1007/978-1-84996-175-2
Springer London Dordrecht Heidelberg New York

British Library Cataloguing in Publication Data
A catalogue record for this book is available from the British Library

Softcover reprint of the hardcover 1st edition 2011

Cover design: eStudio Calamar, Berlin/Figueres

Printed on acid-free paper

Springer is part of Springer Science+Business Media (www.springer.com)

Preface

The IEC Standard for small wind turbine safety, IEC 61400-2, defines a small wind turbine as having a rotor swept area of less than 200 m^2 which corresponds to a rated power of about 50 kW. This approximate definition will be used in this text, which, like the Standard, covers only horizontal-axis wind turbines.

Until the beginning of the twentieth century, all wind turbines were small, at least in terms of power output, and were used for water pumping and milling rather than producing electricity. One of the earliest small turbines for electricity production is shown in Fig. P.1. It was built by English Brothers of Wisbech, England and designed by Edward Burne.

Under circumstances that are not clear, one of Burne's windmills was installed on a farm owned by Russell Grimwade near Frankston, Victoria, Australia, in 1924. Grimwade recorded:

> the electric mains are nowhere within reach. Artificial illumination must be provided and here we displayed our eccentricities to the full. A large [sic] Dutch-type windmill was set up on an attractive hardwood tower that housed the batteries in its base. For artistic effect it gained full marks – for the effective generation of electricity it hardly scored a point. … It was bad engineering that the mill should fail to come up to the wind so that it ran backwards until something broke. …I still believe that man [sic] will someday make use of the power of the wind for his own purpose, and I feel that I have contributed to that research by demonstrating that my method was not the way to do it.[1]

The aim of this book is to demonstrate that, a century later, small wind turbines can be designed and built to avoid many of the problems that faced Grimwade. This is not to say that small turbine technology is mature; there are still areas where it lags well behind current practice for large turbines. This lag is mirrored in the theme of this book which is to provide basic analysis and design guidelines to allow a group of, say, senior engineering undergraduates or junior engineers to design and build a small wind turbine. The approach follows the "Simple Load

[1] pp 141–142 of Poynter JR (1967) Russell Grimwade, Melbourne University Press. Grimwade was technically literate, see for example http://adbonline.anu.edu.au/biogs/A090693b.htm.

Fig. P.1 The Burne small wind turbine on Russell Grimwade's property in the 1920s. Photograph courtesy of the University of Melbourne Archives

Model" (SLM) of IEC 61400-2 which is shown in Chap. 9 to provide straightforward, but necessarily approximate, equations for the main turbine loads and component stresses. There is no equivalent to the SLM in the IEC standard for large turbines.

There are at least five areas where a student or other design group would need additional specialist advice:

- Finite element analysis (FEA) for detailed stress calculations of the critical components
- Electrical engineering advice on the generator and rectifier and possibly the inverter and grid connection
- Detailed dynamics analysis for more accurate stress calculations and fatigue analysis
- Foundation design, and
- Control engineering help in devising and implementing a control strategy.

The first is easily met as FEA is now a standard engineering tool. Its use is highlighted in Chap. 10 on tower design and manufacture. For the second, it is assumed that the turbine's generator will be selected rather than designed and built as part of the project, so the level of knowledge required can be gained from standard texts on the subject. The few issues specific to small turbines are discussed in Chaps. 1, 7 and 11. Detailed dynamics analysis based on "aero-elastic" modeling is still an immature subject for small wind turbines but will undoubtedly develop as more small turbines are built and tested. Some references for aero-elastic modeling are given in the further reading section of Chap. 9. Foundation design is usually site-specific but straightforward once the forces and the base overturning moments are calculated as demonstrated in Chap. 10. There has been a rush of specialist books on wind turbine control and grid interfacing over the last few years, so it would be remiss for this mechanical engineer and aerodynamicist to attempt to match them. Many of the basic control issues are shared by large and small turbines and those that are not are highlighted in the relevant chapters.

Small turbines differ significantly from large ones in blade design and manufacture. The main differences are: low operational Reynolds numbers (*Re*), the need for good low wind performance at even lower *Re*, and the structural requirements of more-rapidly rotating blades. These issues are covered in the first six Chapters and culminate in Chap. 7 on multi-dimensional blade optimisation and manufacture. Most small turbines use "free yaw" whereby a tail fin, rather than a mechanical yaw drive as on larger machines, is used to align the turbine with the wind direction. Yaw behavior and associated issues of tail fin design and aerodynamic over-speed protection are covered in Chap. 8.

The text describes and lists a number of Matlab programs for wind turbine analysis and design. These and supplementary programs, referred to but not listed, can be downloaded from the online material (start at http://extras.springer.com) which also contains additional matter relating to small turbines and the solutions to the Exercises at the end of each chapter. The programs include blade element methods, Chap. 5, multi-dimensional optimisation methods for the design of blades, Chap. 7, and towers, Chap. 10. Excel spreadsheets are provided for noise estimation (Chap. 1) and the loads and component stresses under the IEC Simple Load Model (Chap. 9). All the programs and spreadsheets referred to in the book were written or re-written by the author and have been used for actual turbine analysis and design. The likelihood of errors in them is small but non-zero. They are provided without guarantee. The same applies to the supplementary programs some of which were written by others.

This book is a distillation of more than twenty five years experience working in small wind turbine research, development, and commercialisation. Over the years, my work has been supported by the Australian Research Council, the NSW Renewable Energy Research and Development Fund, the NSW Renewable Energy Development Program, and the Asia-Pacific Partnership on Clean Development. A very important year spent at NASA Ames Research Center was funded by the U.S. National Research Council. There are also many, many people to thank for assistance over that time. I particularly acknowledge Professor Phil Clausen and Paul Peterson who shared much of that time with me. Paul and Sturt Wilson have also shared the vicissitudes of starting and developing a small wind turbine company, Aerogenesis Australia, which incidentally, had its first commercial installation on a farm in Victoria. Sturt Wilson and Phil Clausen provided the FEA of the monopole and lattice tower, respectively, in Chap. 10. Jason Brown wrote the initial version of the SLM spreadsheet in Chap. 9. My graduate students, starting with Phil Clausen and continuing down to Dr. Matthew Clifton-Smith as the last one to complete, have contributed enormously to my knowledge. Most of them appear as co-authors on publications referred to in the main text. I also thank many other colleagues from around the world for providing specific information, answering my questions, listening to my thoughts developing, and correcting them when necessary. Earlier versions of some chapters were used for lecture notes at Newcastle University, where I spent most of those twenty five years, and for a short course at Kathmandu University organised by Dr. Peter Freere. The material was updated and expanded into this text during the first year of my tenure of the ENMAX/Schulich Chair of Renewable Energy at the University of Calgary. I thank the University and the ENMAX Corporation for their vision in supporting distributed generation, here in the form of small wind turbines.

For specific help with this book I thank Peter Freere and Professor Ed Nowicki who co-authored Chap. 11. Phil Clausen and Sturt Wilson gave valuable comments on Chap. 10 and Sturt drew on his blade making skills to improve Chap. 7. Dr. Damien Leqlerq of Cyclopic Energy reviewed Chap. 12 and provided two of the figures. Jim Baxter, Colin Dumais, and Robert Falconer of the ENMAX Corporation provided photographs and information. Colin also brought to my attention several of the interesting web-sites referred to in the book. Mohamed Hammam read the entire manuscript, checked the programs and found and corrected a significant number of typographical errors.

At this point it is customary for authors to thank their family for their supposed forbearance while the book was written. I will not do this because my children have left home and my partner Dr. Cassandra Arnold was working for Medecins Sans Frontieres in Africa for much of that time. However her influence, advice, and proofreading give me much to be thankful for. I also thank my daughter Katie who acquainted me with Burne and Grimwade.

One of my great pleasures over the last twenty five years has been to meet people from around the world who are passionate about small wind turbine technology and its role in mitigating climate change and the huge imbalances in the distribution of wealth and health in this world. I dedicate this book to them and I hope that it will further their efforts. In this regard I acknowledge Springer's generous and enthusiastic agreement to have a special price for the book in developing countries. All royalties from this book will be used to advance the cause of renewable energy in the developing world.

Calgary, May 2011 David Wood

Contents

Symbols and Abbreviations

Because of the wide range of topics covered, a number of symbols have multiple meanings, such as R for blade tip radius and resistance in Ohms. A symbol that has a specific meaning for only one chapter is indicated by giving the chapter number. Many of the symbols used only in Chap. 9 and defined in IEC 61400-2 are not listed here. Table 9.1 lists the present symbols that are different from those in the standard.

Symbols

A	Swept area of blades (m^2)
A	Coefficient in Eq. 4.6
A	Aerofoil cross-sectional area (m^2), Chap. 6
A	Tail fin area (m^2), Chap. 8
A	Cross-sectional area of structural member (m^2), Chap. 10
AR	Aspect ratio
a, a'	Axial and rotational induction factors respectively
a	Weighting factor in evolutionary optimisation, e.g. Eq. 7.2, usually subscripted
a	Side length of polygonal tower (m), Chap. 10
a	Duty cycle, Chap. 11
a_0, a_1, a_2	Coefficients in Eq. 10.6
B	Coefficient in Eq. 4.7
b	Tail fin span (m)
$\mathbf{b}$	Basis vector for evolutionary optimisation, Eq. 7.1
C	Cumulative probability density Eq. 1.18
C	Coefficient in Eq. 4.7
C_a	Axial force coefficient, Eq. 3.10
$C_{a'}$	Tangential force coefficient, Eq. 3.11
C_D	Three-dimensional drag coefficient
C_d	Two-dimensional drag coefficient
C_{d0}	Minimum drag coefficient
C_L	Three-dimensional lift coefficient
C_l	Two-dimensional lift coefficient
$C_{l,\max}$	C_l For maximum lift:drag
C_P	Power coefficient Eq. 1.7
C_P	Aerofoil surface pressure coefficient, Chap. 4
$C_{P,r}$	Extracted power coefficient, Chap. 7
C_Q	Torque coefficient
C_T	Force (thrust) coefficient Eq. 1.13.
c	Blade chord (m)
$\mathbf{c}$	Comparison vector of evolutionary optimisation
D	Drag on three-dimensional body (N)

D	Drag per unit height on a tower (N/m), Chap. 10
d	Drag per unit span on two-dimensional body (N/m)
d	Tower diameter (m), Chap. 10
d	Distance from turbine (m), Chap. 1
d	Distance from rotor to yaw axis (m), Chap. 8
d_0	Tower top diameter (m)
d_1	Slope of linearly-tapered tower
d_h	Tower base diameter (m)
E	Young's modulus (GPa)
e	Eccentricity of rotor centre of mass (m), Chaps. 7 and 9
F	Prandtl tip loss factor, Eq. 5.1
F_y	Yield stress (MPa)
f	Term in Prandtl tip loss factor, Eq. 5.2
g	Acceleration due to gravity = 9.81 m/s^2
H	Effective turbine height (m), Chap. 12
h	Tower height (m)
h_{opt}	Optimum tower height (m), Chap. 12
h_r	Reference height (m), Eqs. 1.14 and 1.15
i	Indent on delta wing, Chap. 8
I	Moment of inertia about the yaw axis (kgm^2), Chap. 8
I	Area moment of inertia (m^2), Chap. 10
I	Current (amps), Chap. 11
I_1, I_2, I_3	Integrals in Eq. 6.10
I_{cp}	Chord-pitch integral Eqs. 6.6, 6.7
I_u	Turbulence intensity, Eq. 1.17
J	Rotational inertia (kg m^2)
K	Lift-slope for a delta wing (1/rad)
K_1, K_2, K_3	Unsteady slender body coefficients, Eq. 8.8
K_{p}, K_{v}	Polhamus coefficients for delta wing, Eq. 8.2
k	Numerical factor in Eq. 8.1
L	Lift on three-dimensional body (N)
L_A	Noise level (dBA), Eq. 1.6
L_p	Sound power level (dB), Eq. 1.4
l	Lift per unit span on two-dimensional body (N/m)
l	Distance from base of tower and turbine centre of mass (m), Chap. 12
M	Moment (Nm)
M_0	Blade root bending moment (Nm)
m	Exponent in power law, Eq. 1.4
m_t	Mass of tower (kg)
m_{tt}	Mass of turbine (tower top mass) (kg)
N	Number of blades
N_{cycles}	Number of fatigue cycles to failure
N_d	Annual average number of lightning strikes, Chap. 11
N_P	Number of poles in generator

N_s	Number of tower sections
n	Number of fatigue cycles
n_1	Structural first natural frequency (Hz), Chap. 10
N_s	Synchronous generator speed (rpm), Chap. 11
P	Power (W)
P	Aerofoil surface pressure (Pa),Chap. 4
P_1, P_2	Pressure on the upwind and downwind face of rotor (Pa), Chap. 2
$\overline{P}$	Average power (W)
p	Probability density function, Eq. 1.19
p	Vortex pitch, Chap. 6
Q	Torque (Nm)
Q	Volume flow rate (m^3/s), Chap. 2
Q_r	Resistive torque (Nm)
R	Blade tip radius (m)
R	Resistance (Ohms), Chap. 11
Re	Reynolds number
r	Radial co-ordinate along blade (m)
r	Distance from tail fin center of pressure to yaw axis (m) (Chap. 8 only)
T	Turbine thrust (N)
T	Temperature (°C)
T	Cable tension (N), Chap. 12
T_d	Turbine design lifetime
T_s	Starting time (s)
t	Time (s)
t	Aerofoil thickness, Chap. 4
t	Tower thickness (m or mm), Chap. 10
t	Trial vector for evolutionary optimisation
U	Wind speed (m/s)
U_p	Wind speed for rated power (m/s)
U_∞	Wind speed in the far-wake (m/s)
V_{tip}	Blade tip circumferential velocity (m/s)
U_{10}	Wind speed at 10 m (m/s)
U_0	Wind speed at hub height (m/s)
U_s	Wind speed for starting (m/s)
U_T	Total velocity at blade element (m/s)
W	Circumferential velocity (m/s)
x	Distance along chord line (m), Chap. 4
x	Tail boom length (m), Chap. 8
Y_1,Y_2	Factors in Eq. 5.6
z_0	Roughness length (m)
α	Coefficient of atmospheric absorption of sound (dB/m), Eq. 1.6
α	Angle of attack (rad)

α_{max}	Angle of attack for maximum lift:drag (rad)
Γ	Circulation (m^2/s)
ζ	Damping ratio
η	Efficiency
θ	Yaw angle (rad)
θp	Blade twist angle (rad)
φ	Wind direction (rad), Chap. 8
λ	Tip speed ratio Eq. 1.10
λ_f	Tip speed ratio at the end of starting
λ_p	Tip speed ratio for rated power
λ_r	Local tip speed ratio, Eq. 3.7
λ_s	Tip speed ratio for starting
μ	Viscosity of air (kg/m/s)
v	Kinematic viscosity of air (m^2/s)
ρ	Density (kg/m^3)
σ	Blade element solidity, Eq. 3.14
σ	Component stress (MPa), Chaps. 9 and 10
σ_a	Axial stress (MPa), Chap. 10
σ_b	Bending stress (MPa), Chap. 10
Ω	Blade speed (usually rad/s)
ϕ	Blade inflow angle (rad), Eq. 3.8
ϕ	Azimuthal angle (rad), Chap. 8
ω	Yaw rate (rad/s)
ω_n	Natural frequency

Subscripts

0	Well upstream of turbine (in undisturbed wind)
1	Upwind face of rotor
2	Downwind face of rotor
∞	Far-wake
b	Blade
design	Design value
LL	Line-to-line
max	Maximum value
P	Rated power
s	Starting
t	Tower
tt	Tower top
tail	Tail
overbar	Time average

Abbreviations

1P	Blade Passing Frequency
AC	Alternating Current
AS	Australian Standard
ASCE	American Society of Civil Engineers
BE	Blade Element
BET	Blade Element Theory
CF	Capacity Factor
CV	Control Volume
DC	Direct Current
IGBT	Insulated Gate Bipolar Transistor
IEC	International Electrotechnical Commission
IEEE	Institute of Electrical and Electronic Engineers
IG	Induction Generator
FEA	Finite Element Analysis
GL	Germanischer Lloyd
KE	Kinetic Energy (J)
MPPT	Maximum Power Point Tracking
NACA	U.S. National Advisory Committee on Aeronautics
NREL	U.S. National Renewable Energy Laboratory
ODE	Ordinary Differential Equation
PD	Power Density (W/m^2)
PMG	Permanent Magnet Generator
RMS	Root Mean Square
SCI	The Steel Construction Institute
SLM	IEC Simple Load Model
THD	Total Harmonic Distortion

Chapter 1
Introduction to Wind Turbine Technology

1.1 How Much Energy is in the Wind?

Since the primary purpose of a wind turbine is to convert the kinetic energy (KE) of the wind into (usually) electrical energy, it is useful to begin by considering the amount of energy and power available, and reviewing the difference between those two concepts. This simple analysis is a gentle introduction to the control volume (CV) analyses that will be used extensively in later chapters.

Suppose the wind is blowing from left to right in Fig. 1.1 with a wind speed of U_0 m/s. For simplicity assume that the wind is steady (i.e. not varying in time) and uniform (i.e. not varying in position). Some effects of unsteadiness (in the form of turbulence) and non-uniformity will be considered later. The air has constant density, ρ, meaning that the flow, as are all flows considered in this book, is incompressible. At 20°C the density of air at sea level is nearly 1.2 kg/m^3; this value can be used in most situations. Most modern turbines are "horizontal-axis" wind turbines, designated as HAWTs, for which the axis of rotation of the blades is parallel or nearly parallel to the wind. Vertical axis wind turbines are not considered in this text.

In Fig. 1.1 the turbine is represented by a circular blade disk whose area $A = \pi R^2$ where R is the blade radius in m. The following analysis determines the kinetic energy in the air that passes the rotor disk per unit time, where the term "rotor" refers to the blades as a set. The analysis is done in the absence of the blades, for reasons that will be explained shortly. The unit of energy is the Joule, J, so the energy that passes will be in J/s, which gives Watts, the unit of power. It is usually power output that concerns the designer and user of wind turbines. However, it is usually electrical energy in the form of kilowatt-hours, kWhs, that is measured and paid for by, say, the electricity utility connected to the turbine.

The right side of Fig. 1.1 shows an elemental volume of the airflow. Its exact shape is not critical. The volume is about to cross the imaginary line (when viewed side-on) in the wind that represents the blade disk. The volume of the element is the product of its area, ΔA, and length normal to the disk, δx, so its mass is $\rho \Delta A \delta x$

D. Wood, *Small Wind Turbines*, Green Energy and Technology,
DOI: 10.1007/978-1-84996-175-2_1,

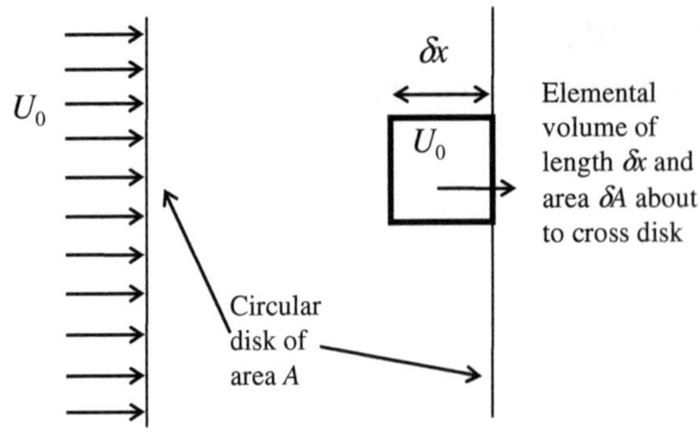

Fig. 1.1 Wind flow past a circular disk representing the blades

and its KE is $\frac{1}{2}\rho\Delta A\delta x U_0^2$. The time taken for this element to cross the blade disk, δt, is given simply by $\delta x = U_0\delta t$. The contribution of the element to the total amount of KE that passes in δt is symbolized as ΔKE, and is given by

$$\delta(\Delta KE) = \frac{1}{2}\rho\Delta A U_0 \delta t U_0^2 \tag{1.1}$$

Summing over all elements of area that make up the disk gives the KE passing the disk as

$$\delta(KE) = \frac{1}{2}\rho A U_0^3 \delta t \tag{1.2}$$

This equation can now be taken formally to the limit as $\delta t \to 0$, to give

$$P = d(KE)/dt = \frac{1}{2}\rho A U_0^3 \tag{1.3}$$

where P is the power, the time rate change (derivative) of the energy. Equation 1.3 is extremely interesting because it suggests, as indeed is approximately the case, that the output power of any turbine depends on the cube of the wind speed.[1] This simple and fundamental fact must never be forgotten. If this cubic dependence seems strange, remember that the wind speed determines both the amount of energy, proportional to U_0^2, and the mass of air carrying that energy through the blade disk per unit time, which is proportional to U_0. In practice the power output is never as great as that suggested by Eq. 1.3 because extraction of all the available KE would require the wind to be decelerated to rest. Furthermore a turbine cannot capture all the wind that would otherwise pass through the disk, even if it could decelerate this flow to rest, so that finding the KE in the absence of the blades will over-estimate the actual energy capture. Including the finite efficiency of the

[1] This result, along with the dependence of power on the rotor area, was established in the 1750s by John Smeaton using a remarkable small-scale experiment. The author has proposed that the approximate scaling of power on rotor area and cube of wind speed be called "Smeaton's Law" in his honour [1], but the proposal has not yet caught on.

drivetrain and the generator, and aerodynamic losses through the action of viscosity, is reasonable to assume that the power converted into electricity is about 40% of that given by (1.3).

It is important that the derivation of (1.1) be understood because the CV analyses that will be undertaken in later chapters extend the ideas and manipulations used to derive (1.1).

Example 1.1 Estimate the power extracted by a 5 m diameter wind turbine at a wind speed of 10 m/s and determine the number of kWhs produced in a day.

Answer From Eq. 1.3 and the discussion following the equation, assume

$$\begin{aligned} P &= 0.4\left(\frac{1}{2}\rho A U_0^3\right) \\ &= 0.4 \times 0.5 \times 1.2 \times \pi \times 2.5^2 \times 10^3 \\ &= 4.71 \times 10^3 \end{aligned}$$

Now check the units: density is in kg/m^3; area in m^2; and (velocity)^3 in m^3/s^3. Their product gives kg m^2/s^3, which are the units of Watts. Thus the estimate for the power output is 4,710 W or 4.71 kW. If the wind speed remained constant over the day, then the number of kWhs produced is $24 \times 4.71 = 113.04$ kWhs per day. Note that the "units" of kWhs per day or month or year are common ways of expresses turbine output and are used despite the fact that they can all be reduced to a multiple of Js.

Example 1.2 Sometimes wind resource surveys give the wind speed in terms of a "power density", *PD*, in W/m^2 which is equal to *P*/*A* from Eq. 1.1. If the power density is 100 W/m^2 what is the wind speed?

Answer

$$PD = \frac{1}{2}\rho U_0^3$$

so that

$$U_0 = (2 \times 100/1.2)^{1/3} = 5.5\,\text{m/s}.$$

1.2 Examples of Wind Turbines

Wind turbines range in power output from a few Watts to tens of megawatts. The IEC safety standard for small wind turbines, IEC 61400-2, defines a small turbine as having a rotor swept area less than 200 m^2, which corresponds roughly to $P < 50$ kW. The precise definition of the boundary between small and large is not critical for this book, so the IEC division is as good as any. The basic operating principles are the same for turbines of all sizes. For example, the restriction on output power given by the Betz–Joukowsky limit, derived in Sect. 2.5, is independent of size. On the other hand, there are operational issues that do depend on size; for example, starting performance and cut-in speed—the lowest wind speed at which power is extracted. Both of these are more important for small machines because:

- Small wind turbines are often located where the power is required or adjacent to the owner's home which may not be the windiest location, whereas wind farms containing large turbines are deliberately sited in windy areas.
- The generators of small turbines often have a significant resistive torque that must be overcome aerodynamically before the blades will start turning. Furthermore, pitch control is rarely used on small wind turbines because of cost. (The precise definition of blade pitch will be given in Chap. 3.) Thus it is not possible to adjust the blade's angle of attack to the prevailing wind conditions. This problem is particularly acute during starting. Starting and low wind speed performance are discussed in Chap. 6.
- Small wind turbine aerodynamics is influenced strongly by low values of the Reynolds number, *Re*. This hugely important parameter is introduced in the next section and its influence on blade aerodynamics is a major topic of Chaps. 4 and 5. Low values of *Re* mean, in practice, that small wind turbines bear greater similarity to model, rather than full-sized, aircraft, and hummingbirds rather than eagles. The later discussion of airfoil lift and drag and blade performance calculations identifies many features particular to small turbines.
- Large wind turbines have complex yaw drive mechanisms to align the rotor to the wind. These are usually deemed too expensive for small turbines, so some form of free yaw is used. The most popular options are to have a tail fin, like three of the five turbines shown in Fig. 1.2, or to have downwind blades, as do the remaining two. Neither choice is optimal for reasons that will be explained in Chap. 8 on tail fin dynamics and yaw behaviour.
- Many small turbines rely on furling for overspeed protection—see Chap. 8—whereas large turbines usually have a brake on the high speed shaft (after the gearbox and before the generator). Aerodynamic overspeed protection is discussed in Chap. 8.

Virtually all large turbines, such as those seen in Fig. 1.3, are upwind machines—the blades are in front of the tower when viewed from the wind direction—and have three blades. The main differences occur in the drivetrain and generator. The most common generator types are doubly fed induction generators (DFIGs) and permanent magnet generators (PMGs), e.g. Burton et al. [2] and Bianchi et al. [3]. DFIGs require a gearbox and are rarely used on small turbines, but PMGs do not. They and the less-used induction generators (IGs) are described in Sect. 1.8 and Chap. 11. There is a much greater diversity of small turbine types as seen in Fig. 1.2 with the number of blades varying from two to seven, and the most popular turbines, the Proven and the Skystream being downwind machines.

Most upwind small turbines have a tail fin which keeps the blades pointing into the wind. The tail fin is designed to minimize θ, the yaw angle of the turbine

Fig. 1.2 A range of small wind turbines. Clockwise from *top left* the Aerogenesis 5 kW turbine, a remote power turbine in Nepal (photo Peter Feere), the Rutland 913 (http://www.marlec.co.uk/), the Proven 15 kW (photo Paul Peterson) and the Southwest Windpower Skystream (photo Jim Baxter)

defined as the angle between the turbine's axis and the wind direction. Yaw reduces the power by a factor of approximately $\cos^2\theta$, e.g. Pedersen [4] and Maeda et al. [5], and so is significant for even moderate values of θ. It is important, therefore, to minimise yaw. Yaw behaviour will be analysed in Chap. 8 along with the associated topics of tailfin design and overspeed protection.

The blades of all wind turbines are comprised of aerofoil sections whose purpose is to produce lift, which is the primary component of the torque about the turbine axis in the direction of blade rotation. For steady flow, the product of this torque and the blade angular velocity, Ω, gives the power extracted from the wind. Blade analysis is introduced in Chap. 3 and the aerodynamics of lift and drag in Chap. 4. The calculation of power output is the subject of Chap. 5. It is not very clear from the photographs in Figs. 1.2 and 1.3 that most turbine blades are twisted, that is they are more "square on" near the tips, but it is more obvious that the blade width, or chord, c, decreases towards the tip. A fuller definition of the twist and chord, along with the reasons why both decrease with radius, are major aspects of wind turbine performance and design, covered in Chap. 5.

Most large turbines have three blades, partly because this number is held to be visually more appealing than the main alternative of two blades. Large blades can be over 60 m long and weigh over 20 tonnes. The small wind turbines in Fig. 1.2 have between two and seven blades. The choice of blade number is a recurrent theme of this book and is discussed in terms of both power extraction (Chaps. 5 and 7) and starting performance, Chaps. 6 and 7. Small blade manufacture and testing is covered in Chap. 7.

1.3 Wind Turbine Noise

In siting a wind turbine, the first and often far from trivial task is to determine the wind resource, which may vary significantly over short distances because of the surface roughness, the topography, and proximity to buildings, trees and the like. These issues are covered in Sect. 1.5 and Chap. 12. There remain at least three further important issues: noise, visual impact, and possible restrictions on tower height. The first two are often addressed for large wind farms using sophisticated software that optimises the layout of the turbines to maximise power extraction and minimise the visual impact of the turbines.

Well designed wind turbines are extremely quiet: one simple data correlation for the sound power level, L_P, gives

$$L_P \approx 10^{-7} P \tag{1.4}$$

[6]; that is one-ten millionth of the turbine's power is output as noise. For this reason, a well designed small wind turbine is almost guaranteed to be quiet. Another correlation that is more accurate in some cases, is

$$L_P \approx 50 \log_{10} \Omega R + 10 \log_{10} R - 1 \tag{1.5}$$

where now L_P is measured in the more common unit of A-weighted decibels (dBA) [7]. Recall that Ω is the blade angular velocity in rad/s, so ΩR is the circumferential velocity of the blade tip in m/s and R is measured in m. L_P is the strength of the source of the sound as a multiple of the standard base level of

10^{-12} Watts. It is used, in combination with an equation for the propagation of the sound, to determine the noise level at any point around the turbine or turbines, e.g. Wagner et al. [7]. The most common "spreading equation" is

$$L_A = L_P - 20\log_{10}\left(\sqrt{h^2 + d^2}\right) - \alpha\sqrt{h^2 + d^2} - 8 \tag{1.6}$$

which is Eq. 10.4.9 of Manwell et al. [8]. It gives the noise level on the ground at distance d from a turbine with a tower of height h. The second term is the hemispherical spreading term which is strictly valid only when $h \gg d$ and the ground is flat. The third term represents the atmospheric absorption of sound with the coefficient α typically in the range 0.002–0.005 dB/m. For small wind turbine purposes, this term is usually negligible. An excel spreadsheet that implements Eqs. 1.5 and 1.6 can be downloaded from online materials for this text by starting at: http://extras.springer.com. Useful guides to turbine noise levels and significance can be found at the web sites listed at the end of the chapter.

Migliore et al. [9] measured the noise output from a number of commercial small turbines. The results are too scattered to attempt to correlate in the simple terms of (1.2 and 1.3). For example, they found the Bergey XL1 1 kW turbine produced so little noise that it was not possible to measure it accurately; a situation in accord to the author's experience with a 5 kW turbine similar to that shown in Fig. 1.3. On the other hand, the 900 W Air 403, whose blades flex to unload the turbine in high winds, had a correspondingly high noise level during the resulting flutter.

Fig. 1.3 Vestas 2 MW V80 wind turbines

1.4 Turbine Operating Parameters

As with any fluid machine, it is often useful to discuss wind turbine operation in terms of parameter groupings that can be obtained from dimensional analysis. Here the important parameters are introduced by taking advantage of Eq. 1.1, which strongly suggests that the most important parameter, the power coefficient, C_P, should be defined as

$$C_P = \frac{P}{\frac{1}{2}\rho U_0^3 \pi R^2} \tag{1.7}$$

C_P the ratio of the actual power produced to the power in the wind that would otherwise pass the blade disk. Note that:

- C_P is dimensionless
- By convention, it includes the factor of ½ to relate power to the kinetic energy flux through the blade disk as determined in Sect. 1.1.
- For later use, note that C_P is not strictly an efficiency, even though it is often treated as one. As will be evident from the next chapter, it is possible to increase C_P by increasing the velocity of the wind through the blades by, for example, surrounding the blades by a diffuser. However, C_P can be interpreted as an efficiency when comparing turbines of the same type, such as the diffuser-less ones considered in this book.

The form of (1.3) helps the dimensional analysis. By making the very general statement that the turbine power should depend on wind speed, air density, turbine radius, Ω, and the kinematic viscosity of the air, ν, then

$$f(P, U_0, \rho, R, \Omega, \nu) = 0 \tag{1.8}$$

where f denotes (the as yet unknown) functional dependence. ν is the actual viscosity divided by the density, and has units of m^2/s. For sea level conditions, $\nu = 1.5 \times 10^{-5}\ m^2/s$ at 20°C. There are many ways of proceeding with the dimensional analysis, all of which should produce the same results. If the following is not familiar, the reader is referred to standard fluid mechanics texts such as White [10].

Equation 1.8 contains six parameters or variables and three dimensions, so there should be three non-dimensional groups resulting from the dimensional analysis. To ensure that C_P as defined by (1.7), is one of these groups, the "repeating variables" must be U_0, ρ, and R. These repeating variables can, in principle, appear in all the non-dimensional groups. Forming these groups then allows (1.8) to be rewritten as

$$f(C_P, \Omega R/U_0, U_0 R/\nu) = 0 \tag{1.9}$$

The second of these groups is sufficiently important to have its own name, the tip speed ratio, and symbol λ. For future reference

$$\lambda = \Omega R/U_0 = V_{\text{tip}}/U_0 \tag{1.10}$$

is the ratio of the circumferential velocity of the blade tips, V_{tip}, to the wind speed. In determining the lift and drag generated on blades in Chap. 4 it will be shown that λ is critical as it sets the angle of attack of the blade sections or elements. Very simply: the tip speed ratio controls the blade aerodynamics. Usually, λ ranges from 7 to 10 for a turbine operating at maximum C_P. Thus the tips are travelling at a velocity that is many times the wind speed and this can cause them to approach the limit of incompressible flow, about 30% of the speed of sound, which is 340 m/s at typical sea level conditions. There is no evidence that compressibility is important on operating wind turbines.

The third group should be recognised as a Reynolds number, *Re*, which generally measures the importance of viscosity. The form of the Reynolds number in (1.9) is *not* used in wind turbine studies. Later consideration of blade aerodynamics will show that the most useful form comes from considering the lift and drag behaviour of the aerofoil sections that comprise the blades. This *Re* contains the blade chord, *c*—the width of the blade—and the "total" velocity at the blade, U_T, both of which vary along the blade. The determination of U_T will be explained in Chap. 3. Thus

$$Re = U_T c/\nu \tag{1.11}$$

It will be seen in Chap. 5 that most of a turbine's power is produced near the tip, so the value of *Re* most often quoted is the tip Reynolds number. If λ is sufficiently high, $U_T \approx \lambda U_0$ at the tip, but $U_T \approx U_0$ everywhere along a stationary or slowly rotating blade. For most large turbines, producing say 100 kW or more, C_P is not strongly dependent on *Re*, but, the next section shows that C_P varies significantly with λ. Reynolds number effects *can* be significant for small turbines, and are considered thoroughly in Chaps. 4 and 5.

The final quantity of interest is the thrust on the blades, *T*. This force is not as important for wind turbines as it is for propellers, which are designed to produce thrust. However, the thrust is usually transmitted to the turbine tower, and so must be included in tower and foundation design. Dimensional analysis starts from the appropriately modified form of (1.9)

$$f(T, U_0, \rho, R, \Omega, \nu) = 0 \tag{1.12}$$

from which the following form of the thrust coefficient, C_T, should be almost immediately recognisable as

$$C_T = \frac{T}{\frac{1}{2}\rho U_0^2 \pi R^2} \tag{1.13}$$

As with the power output, C_T is strongly dependent on λ but not usually on *Re*.

1.5 The Power Curve and the Performance Curve

However much the aerodynamicist is happy to use non-dimensional groups to describe and analyse performance, the owner of the turbine is more interested in the actual power output as function of wind speed. This is given by the power curve. Figure 1.4 shows the power curve for the 7 m diameter Bergey BWC XL 10 kW turbine and Fig. 1.5 shows one for the 80 m diameter Vestas V80 2 MW turbine. Both sets of data are given in Table 1.1. Each figure indicates the "cut-in" wind speed below which no power is produced, approximately 3.5 m/s for both. Also important is the "rated" wind speed where the advertised power is obtained; 13 m/s for the smaller machine and 15 m/s for the larger. It is common for the rated speed to increase with turbine size because the tower height also increases, and this increases the wind speed at the hub. Both turbines have a region, between 5 and 10 m/s, where the power increases rapidly, approximating the cubic dependence of Eq. 1.1. Because of the sensitivity of power to wind speed, it is important to take note of the rated speed; the easiest way for a manufacturer to "improve" the performance of a turbine is to increase the rated speed! This introductory description ignores several important issues in the complex process of determining of the power curve. Some of these will be raised in Sect. 1.7. The interested reader is referred to Measnet [11] for more information.

After 10 m/s, the power output of both turbines increases less rapidly as the control system shifts away from attempting to maximise the power output. It is important to be able to control a turbine at high wind, so that it does not extract more power than can be absorbed by the generator. High power levels may also cause unacceptable structural loads on the blades and other components. Small wind turbine safety is considered in Chaps. 8, 9, and 11.

There are a number of possible control actions for large wind turbines, such as controlling the angle of attack by pitching the blades, that are not available at small

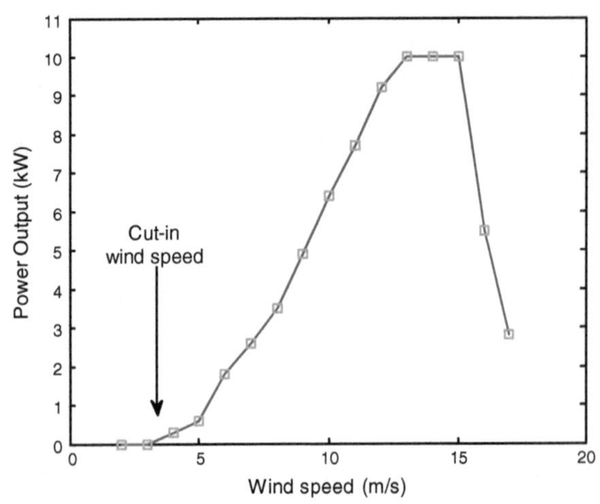

Fig. 1.4 Power Curve for the 10 kW Bergey BWC Turbine at Sea Level. Data from www.retscreen.net (accessed 4 Apr 2010)

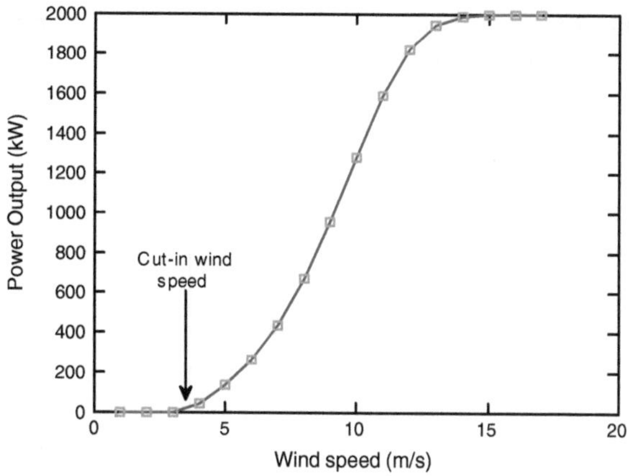

Fig. 1.5 Power curve for the Vestas V80 turbine. Data from www.restscreen.net (accessed 4 Apr 2010)

Table 1.1 Power and power coefficient variation with wind speed for the Vestas V80 2 MW and Bergey XL 10 kW turbines

Wind speed (m/s)	Vestas V80 2 MW		Bergey XL 10 kW	
	P (kW)	C_P	*P* (kW)	C_P
3	0	0	0	0
4	44.0	0.228	0.3	0.203
5	135.0	0.358	0.6	0.208
6	261.0	0.401	1.8	0.361
7	437.0	0.422	2.6	0.328
8	669.0	0.433	3.5	0.296
9	957.9	0.435	4.9	0.291
10	1279.0	0.424	6.4	0.277
11	1590.0	0.396	7.7	0.251
12	1823.0	0.350	9.2	0.231
13	1945.0	0.294	10.0	0.197
14	1988.0	0.240	10.0	0.158
15	2000.0	0.196	10.0	0.128
16	2000.0	0.162	5.5	0.058
17	2000.0	0.135	2.8	0.025

Data from www.retscreen.net (accessed 4 Apr 2010)

scale. For nearly all large turbines but rarely for small ones, there is also a "cut-out" wind speed at which the turbine is shut down for safety reasons. This is 25 m/s for the V80 (not shown in Fig. 1.5). At this speed, the brake is activated, and not released until the wind has died down. At high wind speeds, smaller turbines such as the Bergey 10 kW, are often "furled", that is turned out of the wind direction by the collapse of the tail fin as described in Chap. 8. Other small turbines rely on control of the generator's field current to reduce output in high winds and shorting of the generator output for braking. This "electrical" rather than "mechanical" or "aerodynamic" braking is described in Chap. 11.

As the wind speed increases from below the cut-in, the brake on larger turbines is released, and, the blades are pitched into the wind (this phrase should be clearer

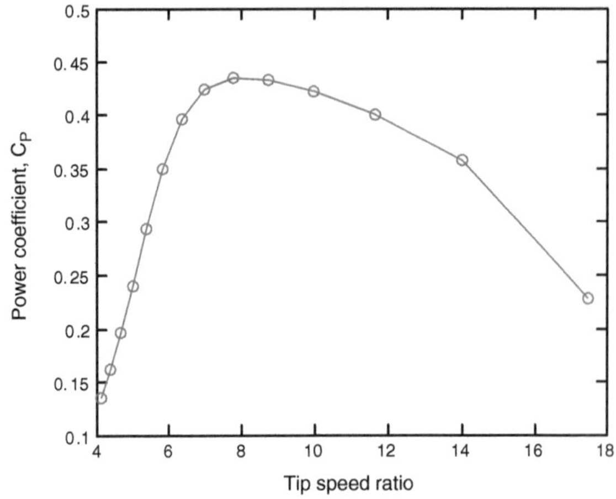

Fig. 1.6 Power coefficient variation with tip speed ratio for the Vestas V80

after Chap. 5) to assist in starting. In contrast, the lack of pitch adjustment on small tribunes makes them rely on the lift generated at high angles of attack, to overcome the resistive torque of the drive train and generator. Some of these aspects will be discussed further in Chap. 6, which highlights the importance of good low wind speed performance of small turbines. The optimisation method described in Chap. 7 is a formal procedure to combine good starting performance with high efficiency of power extraction.

Figure 1.6 shows that λ has a major effect on turbine performance. This is because it controls the angle of attack of the blades. Since the value of C_P determines how much power is extracted at a given wind speed, Fig. 1.6 implies that a constant speed turbine cannot operate at maximum efficiency over a large range of wind speed; obviously the turbine's designers have opted to forgo some of the extractable power for the simplicity of constant speed operation.

Figure 1.6 was generated from the power curve data by using the almost-constant Ω of 16.9 rpm for the Vestas V80. The maximum C_P is nearly 0.44, but note that the power listed in Table 1.1 is the output electrical power which is less than the input aerodynamic power by the product of the efficiencies of the (mechanical) drivetrain and the (electrical) generator. Estimating the combined efficiencies as 0.9, indicates that the V47's maximum efficiency is within 20% of the Betz–Joukowsky [2] limit of $16/27 = 0.593$, the supposed maximum efficiency of this type of turbine (Chap. 2). It is unlikely that any turbine will be able to get significantly closer to the Betz–Joukowsky limit. Figure 1.6 shows the general shape of any performance curve: as λ increases from zero, C_P increases to reach its maximum at the optimal value of λ and then decrease to zero at the "runaway" point where Ω is maximised. For the Vestas V80 the runaway λ is probably around 20. Any turbine that loses its electrical load will accelerate towards runaway. The

[2] This limit is often called the Betz limit. Okulov and van Kuik [12] argued convincingly for the renaming that is followed here.

Table 1.2 Properties of the international standard atmosphere

Altitude (m)	T (°C)	ρ (kg/m^3)	ν (m^2/s)
0	15	1.225	1.46×10^{-5}
1,000	8.5	1.112	1.58×10^{-5}
2,000	2.0	1.007	1.71×10^{-5}
3,000	−4.5	0.909	1.86×10^{-5}
4,000	−11.0	0.819	2.03×10^{-5}
5,000	−17.5	0.738	2.21×10^{-5}
6,000	−24.0	0.606	2.42×10^{-5}

The ISA is available in the Matlab Aerospace blockset and is easily obtained on the internet, e.g. http://www.aerospaceweb.org/design/scripts/atmosphere/

danger is that if the wind speed is high enough, the centrifugal forces within the blades will shatter them. The well-known video clip of a large wind turbine blade shattering, one of the remaining blades then hitting the tower, which buckled and collapsed, is a graphic demonstration of over-speeding.[3] Another example will be discussed in Chap. 12. Figure 1.6 shows why the current trend for both large and small turbines is to attempt to keep a constant λ as the wind speed varies. This is a major challenge for small turbines where the wind speed is not measured. The control strategy to achieve constant-λ operation is an implementation of what is usually called "maximum power point tracking", MPPT. It is explained further in Chap. 12.

The C_P values for the Bergey 10 kW turbine are much lower than for the larger machine. This is due partly to the use of a constant chord, untwisted blade, chosen presumably on the grounds of cost and ease of manufacture, and also to the lower Reynolds numbers which reduces the lift to drag ratio of the blade sections which reduces the torque and hence power, as will be made clear in Chap. 4 (Table 1.2).

Example 1.3 It is proposed to install a remote area power system in a village in Tibet at an altitude of just over 5,000 m. As the RAPS contains a wind turbine, it is necessary to estimate the reduction in power caused by the increase in altitude.

Answer From the material presented in this chapter, it is reasonable to expect that C_P should remain roughly constant as the altitude varies. Thus the change in power occurs through the change in density, much like that shown in Table 1.1 and Fig. 1.6. One way to estimate the density change is from the "International Standard Atmosphere" (ISA) which is often used to account for altitude effects on aircraft performance. Table 1.3 gives the low-altitude ISA variation of temperature, density, and kinematic viscosity with altitude.

The ISA gives the sea-level density of air as 1.225 kg/m^3 (and the temperature as 15°) and the density at 5,000 m as 0.738 kg/m^3. Thus the power reduction

[3] The author first saw this video clip on Youtube. Typing "wind turbine destruction" into a search engine should find it.

Table 1.3 Variation of z_0 and m with terrain from Manwell et al. [8]

Type of terrain	z_0 (mm)	m
Calm open sea	0.2	0.104
Snow	3.0	0.100
Rough pasture	10.0	0.112
Crops	50.0	0.131
Scattered trees	100.0	0.160
Many trees	250.0	0.188
Forest	500.0	0.213
Suburbs	1500.0	0.257
City centres	3000.0	0.289

The values of m are from Eq. 1.16

should be by a factor of 0.738/1.225 = 0.602—a substantial 40%. The corresponding change in the kinematic viscosity is by a factor of nearly 40%. This will cause a similar decrease in the Reynolds numbers, which probably will not have a large effect on optimal power but may influence starting performance.

1.6 The Variation in Wind Speed and Power Output with Height

Example 1.3 showed the sometimes significant effect of the variation in density with altitude and a similar variation will occur if the turbine experiences extremes of temperature. The effects of tower (or nacelle) height, h, however, are more closely associated with vertical variation in the wind speed. Typical values of h are in the range 10–50 m, and are, therefore, small compared to the altitudes (say 5,000 m!) over which the air density and viscosity change significantly.

There are two main expressions used to describe the height dependence of the mean wind speed, which will now be called U rather than U_0 as used previously. The main reason to distinguish between U_0 and U is that $U = U_0$ only when the height, z, is equal to h. The simplest expression for $U(z)$ is the power law

$$U(z) = U(h_r)\left(\frac{z}{h_r}\right)^m \tag{1.14}$$

where h_r denotes a "reference height" usually 10 m, and m is an exponent that depends on the roughness of the surface. It is usually asserted that the logarithmic "law"

$$U(z) = U(h_r)\left(\frac{\ln(z/z_0)}{\ln(h_r/z_0)}\right) \tag{1.15}$$

is more accurate. z_0 is the "roughness length". Typical values of m and z_0 are given in Table 1.3. One possible relation between m and z_0 (in m) is given by Eq. 2.3.28 of Manwell et al. [8] as

$$m = 0.096\ \log_{10}(z_0) + 0.016(\log_{10}(z_0))^2 + 0.24 \tag{1.16}$$

An average value of m is about 1/6, so that, if $P \sim U^3$, then P increases roughly as $h^{1/2}$. This leads to the rule-of-thumb that doubling the tower height will increase the power output by 40%. This approximation ignores the fact that $U(z)$ is the average wind speed and the dependence of the average power output on average wind speed may not be cubic. In fact, the example described in Chap. 12 shows that the dependence is roughly linear. Furthermore, the wind speed dependence on height is often more complex than suggested by the relations (1.14) and (1.15). The logarithmic law, for example, is applicable only if z_0 does not vary significantly around the site for a distance of probably 100 h, the site is flat, and there is no heating or cooling of the air; in other words, the flow is "neutrally buoyant". Thus $U(z)$ may well depend on wind direction and time of day.

1.7 Turbulence and Wind Statistics

Rarely is the wind steady. It usually fluctuates in magnitude, as indicated in Fig. 1.7, and direction in an apparently random manner. The fluctuation level can be measured by the turbulence intensity defined in terms of the root mean square (rms) of the velocity fluctuations. To quantify this relation, assume that the wind speed at any time t, is the sum of the mean U, and the fluctuating velocity $u(t)$. Note that the average value of $u(t)$ is zero. The turbulence intensity is defined as

$$I_u = \frac{\sqrt{\overline{u^2}}}{U} = \frac{1}{U}\left[\frac{1}{T_s}\int_0^{T_s} u^2 dt\right]^{1/2} \tag{1.17}$$

where T_s is the sampling time. The overbar on u^2 (in the square root) denotes a time average—a convention that will be used throughout this book, so that, for example, $\bar{u} = 0$. In practice, the output of an anemometer [which measures $U + u(t)$] is usually sampled at a fixed frequency as in Fig. 1.7 and the integral in (1.17) is approximated by a summation.

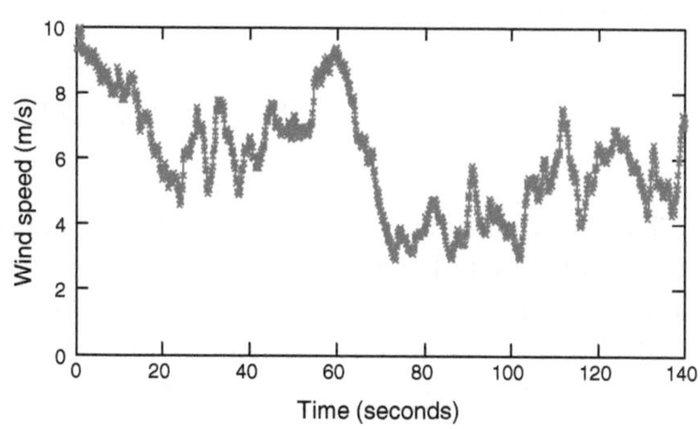

Fig. 1.7 Typical wind speed measured at 10 Hz at the University of Newcastle

The definition (1.17) immediately gives rise to an operational difficulty: what is the appropriate T_s to use in determining I_u? T_s should be sufficiently large that any increase would not alter the value of I_u. In practice, this usually cannot be achieved, and as a compromise, the most common T_s used in wind turbine applications is 10 min. This time is a compromise in that it captures the high frequency fluctuations seen in Fig. 1.7 but misses most of the low frequency changes associated with weather patterns. $T_s = 10$ min is mandated by the International Electrotechnical Commission (IEC) standard for the determination of the power curve—see Measnet [11] for a freely-available summary.

I_u depends on z_0, increasing from around 0.1 (10%) for smooth terrain up to 0.2 (20%) or more for rough terrain at high z_0. Turbulence also depends on height, usually decreasing with increasing h. It can also affect the determination of turbine power—see Exercise 1.23 at the end of this chapter—and influence the turbine loads—see Chap. 9. To determine the average power output from any turbine and to undertake load analysis, it is necessary to know the probability of the wind speed. This probability can be viewed as either the probability density function, $p(U)$, or the cumulative probability, $C(U)$. The former measures the occurrence of a particular wind speed, whereas the latter gives the probability that the wind speed is less than U. Mathematically, the two are related by $dC/dU = p$.

The most common assumption, used, for example, in IEC 61400-2, is the so-called Rayleigh distribution

$$C(U) = 1 - e^{-\pi(U/2\bar{U})^2} \tag{1.18}$$

and

$$p(U) = \frac{\pi U}{2\bar{U}^2} e^{-\pi(U/2\bar{U})^2} \tag{1.19}$$

Note the use of the overbar in (1.18) and (1.19) to denote the average value of U as it is necessary to distinguish between the U, typically found by averaging over 10 min, and its average $\bar{U}$.

The Raleigh distribution is a special case of the Weibull distribution which is commonly used for approximating the wind speed probability distribution. Knowing $p(U)$ allows calculation of the average power output for a particular site, according to

$$\overline{P} = \int_0^\infty P(U)p(U)dU \tag{1.20}$$

This determination is shown in Fig. 1.8 for the power curve in Fig. 1.4 and a mean wind speed of 5 m/s. The ratio of the average power to the rated power is called the capacity factor, a crucial parameter in determining the economics of wind energy systems. For the data in Fig. 1.8 the capacity factor is 0.15. In practice the capacity factor varies widely, reaching nearly 50% at a few very

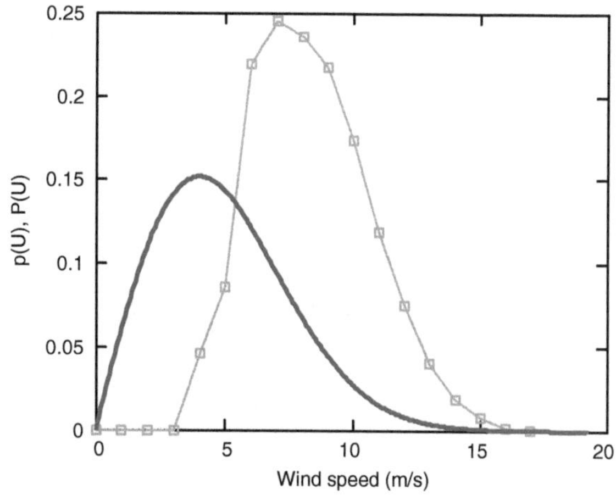

Fig. 1.8 Probability distribution for a mean wind speed of 5 m/s (*solid line*) and the variation of contribution to the power output with wind speed for the Bergey XL 10 kW turbine (*squares*)

favourable locations around the world, and much higher at the Australian Antarctic Base at Mawson where the 10 m average wind speed is 14 m/s. Small wind turbines tend to have much lower capacity factors, and a value of 0.20 is usually regarded as "high". Figure 1.8 also shows another important general fact of wind turbines: the greatest contribution to the average power comes from wind speeds above the mean. If the rated speed is less than about twice the mean speed, then significant reductions in capacity factor are likely. Finally, the broad spread of possible wind speeds at most sites causes the average power output of a turbine to have a different dependence on the average wind speed than the near cubic one exhibited in the power curves of Figs. 1.4 and 1.5.

1.8 The Electrical and Mechanical Layout of Wind Turbines

The generator and bulk of the mechanical components are housed in the nacelle on top of the tower of most turbines. Much of the control electronics is also located on top of the tower for many large and some small ones. For the Vestas V80 in Fig. 1.9 the main component is, of course, the generator, but there are many others, such as a gearbox (#6), which is *not* common on small machines and is slowly losing favour on large machines as well. Most small turbines have little more than the generator in the nacelle, so large turbines are significantly more complex, with, for example, an anemometer and wind vane (#10) used in conjunction with the hydraulic yaw gear (#17) to drive the turbine into the wind. This turbine also has blade pitch adjustment (see #2), as do most large turbines.

The Skystream 2.4 kW turbine is unusual in having much of the elctronics within the nacelle as shown in Figs. 1.10 and 1.11. The bolted shaft in Fig. 1.10 holds the rotor of the permanent magnet generator with the stator attached to the nacelle casting. The blades are attached on the shaft outside the nacelle, which is not shown in the figure. The electronic components comprise the rectifier and

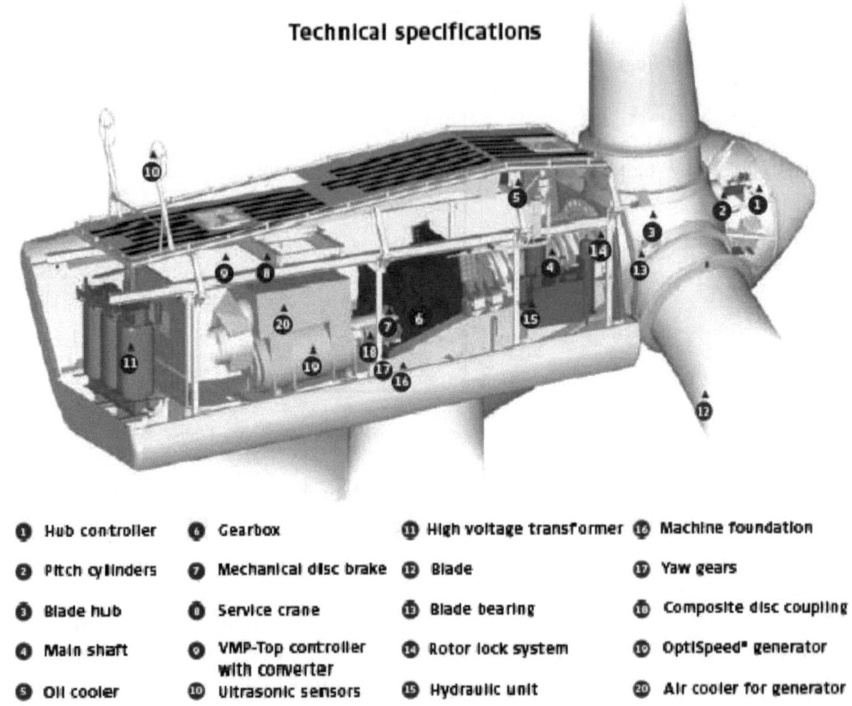

Fig. 1.9 The internal layout of the Vestas V80 2 MW turbine (from product brochure downloaded from http://www.vestas.com/, 10 Apr 2010)

relays for shorting the generator for electrical braking. The Skystream is unusual in that it relies entirely on electronic braking for overspeed control. Figure 1.11 shows the inverter for grid-connection. For battery charging, often as part of a remote power system, a small turbine does not need an inverter. Indeed, most small turbines are not supplied with an inverter.

Pitch adjustment is generally held to be too expensive for small turbines, but some mechanism of overspeed protection is required. Many small turbines are designed to turn the rotor out of the wind in high wind to limit the power—see Exercise 1.2—for protection. This can be done by furling in the horizontal plane or by pitching in the vertical plane. Furling is the more common. The rotor and generator axis is displaced horizontally from the yaw axis (the pivot to the tower) and the resulting yaw moment due to turbine thrust helps to collapse the tail fin at sufficiently high wind speed. There are a number of possible problems with furling, such as:

- The difficulty in achieving furling consistently at the same windspeed for a variety of load cases,
- Furling may cause a non-zero yaw at lower windspeeds with the consequent reduction in power output, and
- The transient forces associated with furling may be large.

Fig. 1.10 Inside the nacelle of the Skystream 2.4 kW wind turbine

Fig. 1.11 The inverter of the Skystream 2.4 kW turbine attached to the nacelle cover

Nevertheless furling is discussed in Chap. 8 along with turbine yaw performance and overspeed protection.

The Skystream and many other small turbines have a permanent magnet generator, PMG. Most of the remainder use standard induction motors. As with many areas of turbine design, there is no unambiguous choice between generator types.

PMGs have the advantages of:

- High efficiency at low rpm and no need for a gearbox
- Simple design
- Ease of control

This is balanced partly by their relatively high cost and the often possibility that the permanent magnet material may lose some of its magnetism over the standard turbine lifetime of say 20 years. Induction generators (IGs) are often common induction motors running "backwards". They have the advantages of:

- Inexpensive due to mass production
- Robust and easy to replace

but are often less efficient especially at small size—see Table 1.4—and part-load. Some of the relative inefficiency is related to the fact that induction generators are not self-exciting whereas PMGs are. The former must be supplied with excitation capacitors which are not guaranteed to work in all situations. For example, if a stationary turbine experiences a very rapid increase in wind-speed, the blade acceleration may be too rapid for excitation to occur. Furthermore, the capacitance may need to be adjusted for changes in cable length between the generator and the controller. Induction generators should be fitted with independent speed encoders for protection and this adds to their cost.

As explained in Chap. 11, basic generator function is independent of size and there are only two important aspects for small wind turbines. The first is the generator inertia which is considered in the next section. Figure 1.12 highlights the second: the "cogging torque" for small PMGs as a fraction of the rated torque (torque at rated power). Cogging torque is required to force the rotor through the stator's permanent magnetic field, e.g. Jahns and Soong [13], and so must be overcome by the blades before they begin to rotate. It depends on the azimuthal angle between the rotor and the stator and its magnitude is independent of the direction of rotation. The data in Fig. 1.12, obtained from the sources in Table 1.5, are the maximum magnitudes of the "cogging torque" and the term will be used with that implication throughout this book.

Note that the cogging torque ratio tends to increase with decreasing PMG output power. Tudorache et al. [14] state that cogging torque must be in the range of 1.5–2.5% of the rated torque for a wind turbine to start at 2.5–3 m/s. Chapters 6

Table 1.4 Typical efficiencies of small induction motors

Motor power (kW)	Number of poles (N_P)	Motor efficiency (%)	Number of poles (N_P)	Motor efficiency (%)	Number of poles (N_P)	Motor efficiency (%)
1.1	4	77.8	6	754	8	72.8
1.5	4	79.2	6	77.5	8	76.5
2.2	4	81.0	6	79.1	8	79.6
3.0	4	82.6	6	81.4	8	82.9

Data from http://www.westernelectric-motors.com/ (accessed 9 Sep 2010)

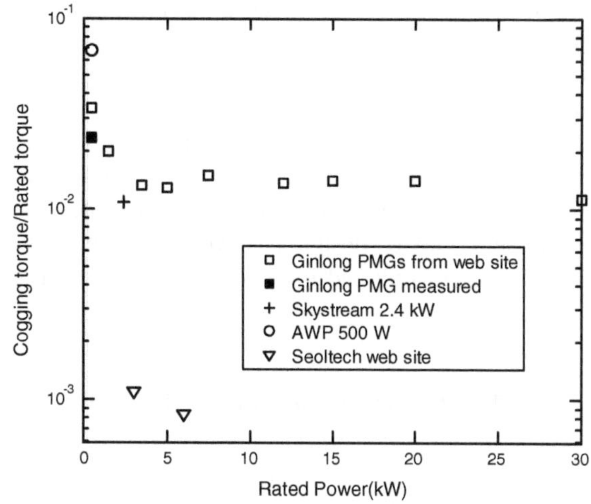

Fig. 1.12 Ratio of cogging to rated torque for a number of small permanent magnet generators. The vertical axis is logarithmic

and 7 analyse the starting and low wind behaviour of small turbines and consider the impact of cogging and other resistive torques. It will be shown that the above estimate of the cogging torque is optimistic: it must be less than 1% of the rated torque to have minimal effect on the starting performance.

On most turbines, the variable frequency and voltage, three phase power from the generator is immediately rectified to a sufficiently high DC voltage which is fed either to a battery bank or to an inverter. An inverter produces mains-frequency voltage A.C. for grid connection or for using standard domestic and other appliances.

What is *not* shown in Figs 1.10 and 1.11 is the controller that regulates the generator and the D.C. voltage to feed the batteries or inverter. Most controllers implement some form of MPPT to attempt to achieve optimal power production as the wind speed varies. With a microprocessor housed in the controller, MMPT often combines a look-up table that gives the maximum power possible at each generator frequency with a strategy to move towards the optimum trajectory if the current power is lower as explained in Chap. 11. Note that frequency and power are easy to measure, the former from the generator frequency, and the latter from the current at the DC bus after the rectifier.

In addition some controllers provide the turbine's primary form of overspeed protection by forcing the blades to operate away from their optimum point (in terms of C_P versus λ) in high winds to limit the output power. In other words, MPPT is modified as the wind speed approaches the rated value to give the shape of the power curves shown in Figs. 1.4 and 1.5. This strategy can work well, but needs some other method for backup in case the load is lost. Typical power electronics and control strategies are covered in Chap. 11, where reference is made to specialist literature for more information.

As small turbines are being increasingly used for grid connection, more and more of them use stand-alone or monopole towers shown in Fig. 1.2. These are similar to towers for large turbines, although they are usually assembled, rather than fabricated, on site. The alternatives of guy-wired tubular towers and lattice towers are still used but the former are mainly for off-grid applications. Guyed towers are often the cheapest, and lattice towers are often the easiest to transport. Tubular lattice towers made from small diameter water pipe are often used where it is not possible to galvanise the tower components. They have the added benefit of being easily and accurately pre-fabricated in short sections using simple equipment. Tower design is covered in Chap. 10 and the important issues in installation, raising and lowering are discussed in Chap. 12.

1.9 The Size Dependence of Turbine Parameters

IEC 61400-2 defines a small turbine as having a swept area of less than 200 m^2, which corresponds to a power output of about 50 kW. This definition is as arbitrary as it is necessary. There is a further subdivision in that turbines of less than 2 m^2 (about 500 W) do not need to have their tower included in the certification process. Clausen and Wood [15] have made a further subdivision as shown in Table 1.6.

The significance of the division is more apparent from Table 1.6 which shows the size-dependence of the main turbine parameters. The scaling of the centrifugal loads follows from the need to keep λ independent of turbine size. The dependence of the noise follows from Eq. 1.4, and that of the power should be obvious by now. From Exercise 1.4, all torques, including the starting torque scale as the cube of the radius. The last two scalings, of blade mass and inertia, follow from assuming that the blade density and shape stay constant as turbine size changes. Figure 1.13 shows the actual power law dependence is not as strong as indicated in Table 1.7 for large blades made by LM Glassfiber. Brøndsted et al. [16] found an exponent closer to 2.66 for, presumably, a different set of blades. It is likely that improvements in manufacturing techniques have reduced blade mass and lead to an effective reduction in the exponent; the longer the blade the more recent it is likely to be.

Table 1.5 Source for data in Fig. 1.12

Label	Source	Comments
Ginlong (a)	www.ginlong.com (accessed 6 July 2010)	Web site gives maximum values
Ginlong (b)	Measured by author	Value the same on two units
Skystream	Measured by author on one turbine only	www.skystreamenergy.com
AWP[a]	Measured by Wright and Wood [17]	This 500 W turbine features significantly in Chaps. 8 and 9
Seoltech	www.skywindearth.com (accessed 15 Sep 2010)	Web site gives maximum values

[a] AWP stands for Australian Windpower which is no longer in production

Table 1.6 Typical small turbine operating parameters

Category	P (kW)	R (m)	Max. Ω (rpm)	Typical uses	Generator type(s)
Micro	≤ 0.5	1.5	700	Electric fences yachts	Permanent magnet (PMG)
Mid-range	0.5–5	2.5	400	Remote power systems, single-user grid connections	PMG or induction
Mini	20–50+	5	200	Mini grids, remote communities	PMG or induction

Adapted from Clausen and Wood [14]

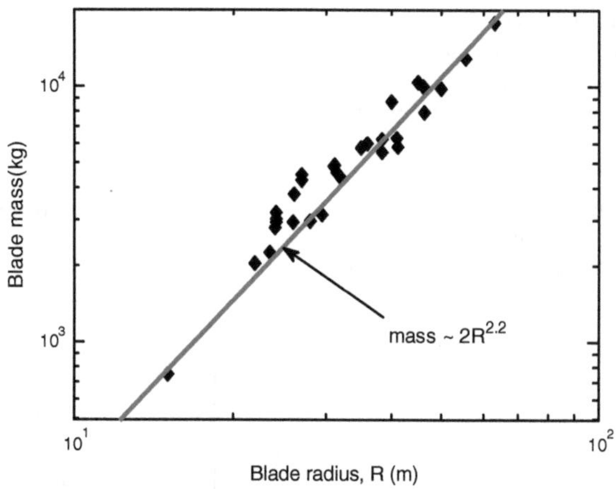

Fig. 1.13 Variation of blade mass with radius for the LM range. Data downloaded from www.lmglassfiber.com (accessed 19 May 2010)

Because of their high operational speed and low torque from Table 1.7, micro-turbines have the poorest starting performance which is often exacerbated by PMG cogging torque. A related and size-dependent issue is the generator inertia, J. It will be shown in Chap. 6 that generator inertia is important for two reasons. First, if it and the cogging torque can be neglected, then turbine starting becomes independent of the number of blades. Secondly, a generator with inertia large compared to that of the blades it will also impair starting. Figure 1.14 shows some data on J for small generators plus a line of fit that suggests $J \sim P^{3/2} \sim R^3$. The IG data does *not* include gearbox inertia which should also scale on P. This neglect is partly mitigated by showing J only for 8-pole IGs which would require a smaller gearbox than the lower-pole IGs in Table 1.4. Since blade inertia scales on a larger power of R, it can be concluded that, if the generator inertia of micro-turbines can be neglected—as will be demonstrated in Chap. 6—then it is safe to exclude generator inertia from starting considerations for turbines of all sizes.

A further interesting fact of the sub-division of small turbine sizes is that micro-turbines are by far the most common small turbine. They are mass produced in China, Britain, and the USA, and the number sold vastly outstrips the number of the mid-range and mini turbines, only some of which are mass produced.

The remaining issue of turbine size to be discussed is the often critical one of cost. Large wind turbines cost typically around $US 2 per watt of installed

Table 1.7 Dependence of main turbine parameters on blade radius, adapted from Wood [16]

Parameter	Power law dependence on *R*
Blade angular velocity	−1
Centrifugal blade loads	−1
Reynolds number	1
Gearbox ratio	1
Power output	2
Noise output	2
Brake torque (high speed side)	2
Brake torque (low speed side)	3
Starting torque	3
Blade mass	2–3
Gyroscopic moments	3–4
Inertia of blades	4–5

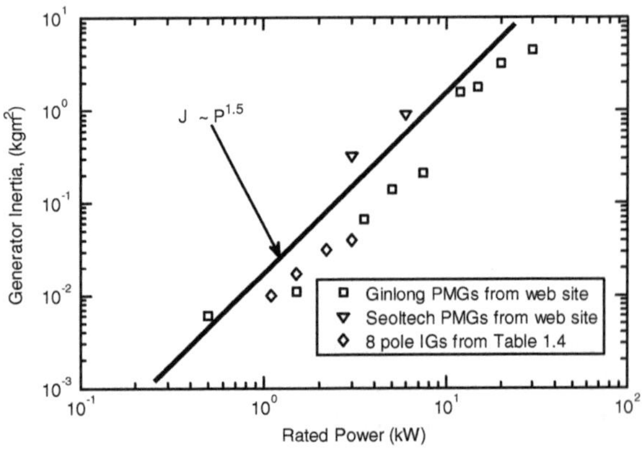

Fig. 1.14 Generator inertia versus rated power for a number of small generators

generating capacity. Currently the cheapest small turbines cost about $US 5 per watt. The high cost of all small scale renewable technologies has lead to the advent of feed-in tariffs in many Western countries, which has driven the recent rapid expansion of the small wind turbine market. Feed-in tariffs are a preferential high price legislated for small scale renewable generators, e.g. Wood [18].

A typical breakdown of costs for a small turbine is given in Table 1.8. They are very similar to those shown in Table 9.1 of Burton et al. [2] for a 10 MW wind farm.

1.9.1 Further Reading

There are a number of good introductory books on wind turbines, such as Kentfield [19] with a chapter on tail fin furling, Manwell et al. [8] and Spera [20] which contains an excellent introduction to aerodynamics by R. E. Wilson. An old

Table 1.8 The percentage installed costs for the Aerogenesis 5 kW turbine shown in Fig. 1.2

Component	Percentage of total cost
Blades	7
Platform, tail fin	5
Gearbox, generator, brake	6
Nose cone and cover	3
Controller/inverter	18
Tower	32
Installation and grid connection	29

favourite, but no longer in print, is Eggleston and Stoddard [21] with its detailed treatment of blade and turbine dynamics. The more recent book on large turbines by Burton et al. [2] is probably the closest in scope to the present.

Wind energy is fortunate in having a number of excellent web sites, such as the Danish Wind Turbine Manufacturers site: http://www.windpower.org/. Related information is available from the American Wind Energy Association site: http://awea.org along with a list of, and contact details for, American manufacturers of small turbines. The British Wind Energy Association, now known as Renewable UK: http://www.bwea.com/ has a lot of information on small turbine issues. The U.S. National Renewable Energy Laboratory (NREL) has probably done more than any other organisation to promote and develop small wind energy technology. To reach their extensive range of publically-available reports on small wind turbine testing and development go to http://www.nrel.gov/wind/pubs_research.html#turbine

The Nordic Folkecenter for Renewable Energy in Denmark publishes an annual catalogue of small wind turbines. Further information from info@folkecenter.dk. The German Wind Energy Association (BWE) produces a similar market survey for large and small turbines available from: http://www.wind-energie.de/de/publikationen/ which is also highly recommended.

The IEC standards covering most aspects of large and small wind turbine design and safety, performance evaluation can be found at www.iec.ch. However, the standards are expensive!

Readers intending to proceed to the subsequent chapters are warned that their knowledge of basic fluid mechanics will be tested. If you wish to revise, or to learn, the necessary background material, then you should consult an introductory text, such as White [10].

1.9.2 Exercises

1. In the Preface and at the start of Sect. 1.2 it was stated that a rotor swept area of 200 m^2 corresponds to a rated power of about 50 kW. Taking the rated wind speed as 10 m/s, check whether that statement is valid.
2. Check the units of Example 1.2.

3. The maximum solar insolation (radiant energy from the sun) that reaches the earth's surface is about 1 kW/m^2. In terms of the power density used in Example 1.2, what wind speed corresponds to this level?
4. If the shaft torque on a turbine is denoted as Q Nm, show that the appropriate formulation of the torque coefficient, is

$$C_Q = \frac{Q}{\frac{1}{2}\rho U_0^2 \pi R^3}$$

5. Determine the relationship between C_Q and C_P: $C_P = \lambda\, C_Q$
6. Show that V_{tip} for the Vestas V80 2 MW turbine is 70.0 m/s and that the sound power level using Eq. 1.4 is 107.3 dB.
7. A 5 m diameter 5 kW turbine has a shaft speed of 280 r.p.m. at a wind speed of 10 m/s. Determine V_{tip} and estimate the sound power level from Eq. 1.4.
8. The wing of a Boeing 747 has a tip chord of 4.03 m. What is the tip *Re* when:

   (a) taking-off at sea-level with a speed of 84 m/s.
   (b) cruising at 240 m/s at an altitude where $\nu = 3.2 \times 10^{-6}$ m^2/s.

9. The Vestas V80 turbine whose characteristics are shown in Table 1.1 and Figs. 1.5 and 1.6 has a tip chord of 0.48 m. Estimate the tip *Re* at $U = 5$ and 10 m/s.
10. The blades of a 500 W turbine have a tip chord of 42 mm and is designed to start at $U_0 = 3$ m/s. Show that the tip *Re* for the stationary blades at this wind speed is 8,600.
11. Estimate the tip *Re* for the turbine in the previous exercise if its rated speed is 10 m/s at which $\lambda = 7.5$.
12. If the turbine for the previous exercise has a tip radius of 0.97 m, and it starts at a wind speed below 3 m/s, what power would it extract at 3 m/s?
13. The blades of a 5 m diameter 5 kW turbine have a tip chord of 75 mm. What is the *Re* at rated conditions, $\lambda = 10$, and $U_0 = 10$ m/s. If the speed of sound is 342 m/s at sea level conditions, what is the tip Mach number.
14. What additional non-dimensional group would arise if blade chord had been added to the general equation for power or thrust?
15. Table 1.9 gives the rotor diameter of some Vestas turbines as a function of output power. Show that there is no discernible change in efficiency as rotor size increases.

Table 1.9 Power, rotor diameter, and rated speed for a range of Vestas turbines

P (kW)	Rotor diameter (m)	Rated speed (m/s)
600	39	18
660	47	17
850	65	17
1650	66	18
3000	90	15

Data from www.retscreen.net (accessed 20 Jun 2010)

16. Consider a two-bladed 5 m diameter turbine with a hub height of 20 m. Using (1.15), estimate the difference in wind speed at the tips of the blades when they are vertical for the smallest and largest values of z_0 in Table 1.3. What is the significance of these results?
17. Assume that $P \sim U^3$ and the cost of a turbine is a (in an arbitrary currency), and the cost of a tower in the same currency is bh^n, where a, b, and n are "constants" and h is the hub (and tower) height. Further assume that the variation in U across the blade disk (see the previous exercise) can be ignored. Derive an expression for the power per unit cost, $p = P/(a + bh^n)$, as a function of h, using Eq. 1.14 for the variation of U with h. Defining r as the ratio of turbine cost to the cost of the turbine and tower, $r = a/(a + bh^n)$, show that the optimum tower height h_{opt}, that maximises p, occurs when $r(h_{opt}) = 3m/n$, provided $3m/n < 1$. Note that m is the power law exponent in Eq. 1.14.
18. Check the websites of some small turbine and tower manufacturers to obtain data for a, b, and n in the previous exercise. See whether the optimum height increases with increasing roughness, and if there is an optimum height for the roughest terrain in Table 1.3.
19. By considering the power curves in Figs. 1.4 and 1.5, comment on the use of $P \sim U^3$ in the previous two exercises. Using manufacturer's data such as in Table 1.1 determine an alternative exponent for $P \sim U^l$ and hence show $r(h_{opt}) = lm/n$, provided $lm/n < 1$.
20. Using the logarithmic law, (1.15), show that r defined in Exercise 17 is now given by the implicit equation, $r = 3/[n \ln(h_{opt}/z_0)]$ with no restriction on n and z_0 so that there is always an optimum height.
21. Writing $U(t) = U + u(t)$, and assuming that C_P in (1.7) is independent of t and $U(t)$, and

$$\overline{u^3}/U^3 \ll 3I_u^2$$

show that

$$\overline{P} \approx \frac{1}{2}\rho U^3 \pi R_1^2 C_p \left(1 + 3I_u^2\right)$$

In words: determining C_P from the power curve by ignoring the turbulence, as was done in producing Fig. 1.6 and Table 1.1, will produce an over-estimate. If $I_u = 0.2$, the error is 12%. Comment on the empirical fact that the power actually increases as the sampling period is reduced from the standard value of 10 min.
22. From the Rayleigh distribution, (1.13) with $\bar{U} = 6$, 8, and 10 m/s, calculate the capacity factor for the Vestas V80 turbine using the data in Table 1.1.
23. From the Rayleigh distribution, (1.13) with $\bar{U} = 6$, 8, and 10 m/s, calculate the capacity factor for the Bergey XL 10 kW turbine using the data in Table 1.1.

24. As noted in the text, the Skystream 2.4 kW turbine has all its electronics in the nacelle, rather than mounted in the tower at its base or a nearby special container. What are the advantages and disadvantages of these possible locations for the electronics?
25. Starting with the web sites listed in Table 1.9, find the typical efficiencies of PMG in the range up to 10 kW to see whether they are more efficient than the IGs in Table 1.4.
26. The "theoretical" n for the R^n-dependence of blade mass is 3, which is greater than the actual value seen in Fig. 1.13. Why would you expect this behaviour, rather than the actual value being greater than 3?
27. The blades on the V80 turbine weigh 6,500 kg each. Compare this to the data correlation of blade mass in Fig. 1.13.
28. If the blade mass scaled as the cube of the radius, what would be the mass of a 2.5 m blade given that the 40 m blade weighs 6,500 kg. Answer: 1.59 kg
29. A standard induction motor has a synchronous speed given by $n_s = 120f/N_P$ where f is the frequency (in Hz) and N_P is the number of poles. If the maximum f is limited to 50 Hz, and ignoring generator slip, determine whether it is possible to design a small wind turbine for each of the generators listed in Table 1.5 *without* the use of a gearbox.

References

1. Wood DH (2000) Smeaton's law? Wind Eng 23:373–375
2. Burton T, Sharpe D, Jenkins N, Bossanyi E (2001) Wind energy handbook. Wiley, Chichester
3. Bianchi FD, de Battista H, Mantz RJ (2007) Wind turbine control systems. Springer, London
4. Pedersen T (2004) On wind turbine power performance measurements at inclined airflow. Wind Energy 7:163–174
5. Maeda T, Kamada Y, Suzuki J, Fujioka H (2008) Performance under yawed inflow. J Solar Energy Eng 130:1–7
6. Lowson M (1992) Applications of aero-acoustic analysis to wind turbine noise control. Wind Eng 16:126–140
7. Wagner S, Bareiss R, Guidati G (1996) Wind turbine noise. Springer, Berlin
8. Manwell JF, Mcgowan JG, Rogers AL (2002) Wind energy explained—theory, design and application. Wiley, New York
9. Migliore P, van Dam J, Huskey A (2004) Acoustic tests of small wind turbines. In: AIAA Paper 2004-1185
10. White F (2011) Fluid mechanics, 7th edn. McGraw Hill, New York
11. Measnet (2009) Power performance measurement: version 5. http://www.measnet.com/power5.pdf. (accessed 15 Oct 2010)
12. Okulov, VL, van Kuik GAM (2011) The Betz–Joukowsky limit: on the contribution to rotor aerodynamics by the British, German and Russian scientific schools. Wind Energy. doi: 10.1002/we.464
13. Jahns TM, Soong WL (1996) Pulsating torque minimization techniques for permanent magnet AC motor drives—a review. IEEE Trans Ind Electron 43:321–330

14. Tudorache T, Melcescu L, Popescu M (2010) Methods for cogging torque reduction of directly driven PM wind generators, 12th International Conference Optimization of Electrical and Electronic Equipment (OPTIM). doi:10.1109/OPTIM.2010.5510390
15. Clausen PD, Wood DH (2000) Recent advances in small wind turbine technology. Wind Eng 24:189–201
16. Brøndsted P, Lilholt H, Lystrup A (2005) Composite materials for wind power turbine blades. Ann Rev Mater Res 35:505–538
17. Wright AD, Wood DH (2004) The starting and low wind speed behaviour of a small horizontal axis wind turbine. J Wind Eng Ind Aerodyn 92:1265–1279
18. Wood DH (2010) Small wind turbines for remote power and distributed generation. Wind Eng 34:241–254
19. Kentfield JAC (1996) The fundamentals of wind-driven water pumpers. Gordon & Breach, Amsterdam
20. Spera D (1994) Wind turbine technology. ASME Publishing, New York
21. Eggleston DM, Stoddard FS (1987) Wind turbine engineering design. van Nostrand, New York

Chapter 2
Control Volume Analysis for Wind Turbines

2.1 Introduction

This chapter uses the control volume (CV) method introduced informally in Sect. 1.1, to develop the basic equations for conservation of mass (Sect. 2.3), momentum (2.4), angular momentum (2.5), and energy (2.6). The forms of the equations to be used for this finite CV are derived in any standard undergraduate textbook on fluid mechanics such as [1]. It must be remembered that the basic analysis undertaken here (and up to the end of Chap. 4) assumes steady, uniform flow—conditions that rarely occur in any real wind! This chapter makes the further assumption that the flow through the blades is radially uniform, that is, the velocities (and therefore pressures) do not vary with radius.

In Sect. 2.7, these equations are used in conjunction with the turbine operating parameters to determine the Betz–Joukowsky limit, the theoretical upper limit on the power extraction capability of a wind turbine.

2.2 The Control Volume

Figure 2.1 shows the CV to analysis wind turbine behaviour. The CV is a cylinder whose radius, R_{CV}, is very much larger than the blade tip radius, R. The upstream face of the CV (at the left-hand side) is located far enough upstream that the velocity entering the CV is the wind speed U_0 and the pressure is ambient or zero gauge pressure. In other words, the presence of the blades does not influence the flow entering the upstream face of the CV. Because the turbine extracts energy from the wind, the velocity in the far-wake, U_∞, is always less than U_0. The radius of the far-wake is R_∞. Thus the wake expands as shown by the "bounding streamline" which is part of the "bounding streamtube", the boundary between the flow passing through the blades and the "external" flow. This boundary is sharp and can support a discontinuity in the velocity and pressure across the bounding streamline. It will be shown in Chap. 5 that the bounding streamtube behind the

D. Wood, *Small Wind Turbines*, Green Energy and Technology,
DOI: 10.1007/978-1-84996-175-2_2, © Springer-Verlag London Limited 2011

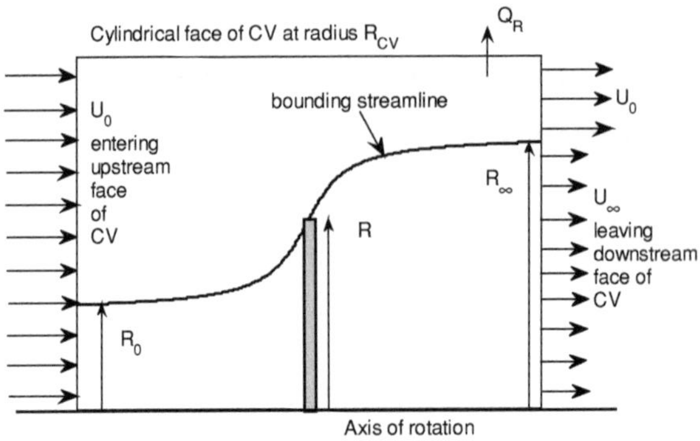

Fig. 2.1 Control volume for wind turbine of radius R in steady uniform flow

rotor is comprised of helical vortices trailed from the blade tips in much the same way that nearly straight tip vortices are shed at the tips of aircraft wings. The key difference between helical and straight vortices is that the former can induce an axial flow whereas the latter cannot. Trailing vortices are a consequence of Kelvin's theorem that circulation must be continuous in an otherwise inviscid fluid, so the "bound" vorticity of turbine blades and aircraft wings must be shed into their wakes. For blades, this shedding occurs at the hubs as well as the tips but the hub vorticity does not appear to have a leading-order effect on the flow.

The figure shows that there is expansion of the flow *before* the blades; in fact, about one-half the expansion, as measured by the cross-sectional area of the bounding streamtube, occurs in the upstream flow. This is one reason why the turbine can never convert all the kinetic energy that would pass through the blade area in the absence of the blades.

It is further assumed that U_∞ and the pressure in the far-wake are uniform, and that the latter is equal to atmospheric pressure. Furthermore, the presence of any swirl, or circumferential velocity generated by the blades, is ignored, even though the torque on the blades must result in a change in the angular momentum of the air. The accuracy of this assumption is examined in Chap. 4. Briefly, for normal wind turbine values of the tip speed ratio, λ, the circumferential velocity is so low that it can be neglected when considering conservation of momentum and energy, and, in any case, it does not enter the conservation of mass equation.

The three conservation equations for an incompressible airflow (constant density) are now applied by assuming that the flow is uniform and steady, which means that there is no accumulation of mass, momentum, angular momentum, and energy within the CV.

2.3 Conservation of Mass

When divided through by the constant density, ρ, the vector form of the conservation of mass equation for a steady flow is

$$\int \mathbf{U} \cdot d\mathbf{A} = 0 \tag{2.1}$$

where $d\mathbf{A}$ indicates a vector element of the area on the CV faces. $d\mathbf{A}$ is, by convention, always pointing outwards (at right angles) from the CV. This direction is called the "outward facing normal". It is important to remember this convention and its critical use in determining the signs of the contributions from each face. Note also that, as with all conservation equations for steady flow, the only terms in (2.1) come from the CV faces. This is because no conserved quantity can accumulate within the CV for a steady flow. In words: at every instant, the amount of air entering the CV per unit time must be balanced by the same amount leaving from a different part of the CV, again per unit time. (The molecules comprising these amounts are, of course, different. If you have trouble with this concept, think of the water entering a hose from the tap, with the same amount leaving the end of the hose. The molecules leaving at any time are *not* those entering at the same time.)

For the above CV, air enters from the upstream face, causing a negative contribution to (2.1), because $d\mathbf{A}$ is in the direction opposite to $\mathbf{U}$, and leaves from the downstream face in the far-wake, giving a positive contribution, as then $d\mathbf{A}$ is in the same direction as $\mathbf{U}$. There is also a positive contribution from the cylindrical face at radius R_{cv}.

At the upstream face, the magnitude of the velocity is constant and equal to U_0. To reiterate: this velocity is in the opposite direction to the outward pointing normal, so that $\mathbf{U}{\cdot}d\mathbf{A}$ will be negative and have the value of $-U_0 dA$ where dA is now a scalar element of area. Thus the contribution to the integral in (2.1) is $-U_0 \pi R_{cv}^2$. (Note that the result of a vector dot product is a scalar.) Similarly, the contribution from the face in the far-wake is $U_0\pi(R_{cv}^2 - R_\infty^2) + U_\infty \pi R_\infty^2$. All these terms have the units of *velocity* × *area* or m^3/s, and are usually termed "volume flow rates" because they give the volume of air that passes the particular face every second. Usually volume flow rates are given the symbol Q, but this symbol is used in this text for torque. A Q with a subscript will represent a volume flow rate for this and the next sections only. If Q_R represents the flow rate out of the cylindrical face of the CV, then Eq. 2.1 gives

$$-U_0\pi R^2 + U_0\pi\left(R_{cv}^2 - R_\infty^2\right) + U_\infty \pi R_\infty^2 + Q_R = 0$$

or

$$Q_R = (U_0 - U_\infty)\pi R_\infty^2 \tag{2.2}$$

Q_R must be due to a radial velocity. The average value of that velocity, V_R, multiplied by the flow area, will equal Q_R. If the length (in the wind direction) of the CV is X, say, the flow area is $2\pi R_{CV} X$, so V_R can be made arbitrarily small by increasing R_{CV}. In fact, the following analysis requires $R_{CV} \gg R$, in order to make V_R negligible and Q_R independent of R_{CV}.

Another important use of the conservation of mass equation is to fix the volume flow rate within the bounding streamtube. By an appropriate change to the CV shown in the figure, it is easy to deduce that

$$Q_0 = U_0 \pi R_0^2 = Q_1 = U_1 \pi R_1^2 = Q_\infty = U_\infty \pi R_\infty^2 \tag{2.3}$$

so that the volume flow rate within the bounding streamtube at any axial location in the flow, is constant.

2.4 Conservation of Momentum

Newton's law in CV form determines the force acting *on* the air, which is the negative of the force (thrust) acting on the blades, **T**, in vector form. Thus the equation for **T** is

$$\mathbf{T} = -\rho \int \mathbf{U}\mathbf{U} \cdot d\mathbf{A} \tag{2.4}$$

Focusing only on T which is the force in the direction of the wind, it is easier to revert to the scalar component of (2.4) in the direction of the wind. The pressure is constant and equal at all CV faces, so it does not contribute to the momentum equation, as has already been assumed in writing (2.4). Furthermore, the velocities at the CV faces are uniform (even if the velocity at the downstream face is discontinuous at R_∞) so the application of Eq. 2.4 proceeds in the same manner as for Eq. 2.1. The result is

$$T = \rho U_0 U_0 \pi R_{cv}^2 - \rho U_0 U_0 \pi \left(R_{cv}^2 - R_\infty^2\right) - \rho U_\infty U_\infty \pi R_\infty^2 - \rho U_0 Q_R \tag{2.5a}$$

and is, therefore, positive in the direction of the wind. The most interesting term in this equation is the last, representing the removal of momentum (equal to U_0 per unit mass) by the volume flow rate (Q_R) out of the cylindrical face of the CV. Using Eq. 2.2 this term can be removed and (2.5a) rewritten as

$$T = \rho U_\infty \pi R_\infty^2 (U_0 - U_\infty) = \rho Q_\infty (U_0 - U_\infty) \tag{2.5b}$$

Another equation can be derived for T by considering the flow through the "disk" representing the rotating blades. Imagine that the blades can be replaced by a thin, uniform circular disk across which the velocity is continuous but the pressure is discontinuous, then T can result only from the pressure difference $P_1 - P_2$. P_1 acts in the wind direction on the upwind side of the disk and P_2 acts upwind on the downwind side. (Note that the symbol P is used both for power when there is no subscript and pressure when it is subscripted.) Idealising the blades as an infinitely thin porous disk—often called an "actuator disk"—is a common one in the analysis of fluid machines. It can be thought of as a model for a rotor with an infinite number of infinitely thin blades. Thus

$$T = (P_1 - P_2)\pi R^2 \tag{2.6}$$

2.5 Conservation of Angular Momentum

The torque on the blades is equal and opposite to that acting on the air. The equation for the vector torque **Q** is

$$\mathbf{Q} = -\rho \int \mathbf{r} \times \mathbf{U}\mathbf{U} \cdot d\mathbf{A} \tag{2.7}$$

If there is no swirl (or angular momentum) in the upstream flow, the only contribution to (2.7) comes from the CV face in the far-wake. Equation 2.7 can thus be turned into a scalar equation for the contribution to Q, the torque acting around the axis of rotation, which is normally the only torque of interest. To do this recognise that the magnitude of $\mathbf{r} \times \mathbf{U}$ is rW_∞, where W_∞ is the swirl velocity (about the turbine axis) in the far-wake. Furthermore, rW_∞ is related to a very important quantity called the circulation around each blade, Γ, by

$$N\Gamma = 2\pi r_\infty W_\infty \tag{2.8}$$

where N is the number of blades. It is shown in Chap. 4 that Γ is nearly constant in the far-wake. Substitution into (2.7) leads to the following equation for C_Q, the torque coefficient

$$C_Q = \frac{Q}{\frac{1}{2}\rho U_0^2 \pi R^3} = \frac{N\Gamma U_1}{\pi} \tag{2.9}$$

The torque is related to the power by

$$P = Q\Omega \tag{2.10}$$

and is imparted to the blades by the aerodynamic forces (principally lift and drag) generated by the flow through the blades.

2.6 Conservation of Energy

Finally, consider the energy equation for the CV used in the application of the mass and momentum conservation equations. To start, recall the assumption that the pressure in the far-wake is atmospheric (zero gauge pressure) so the pressure on all faces of the CV is atmospheric. This means no net work is done by the pressure forces in moving fluid into or out of the CV, so the only form of energy that to be considered (in the ideal case) is kinetic energy. The conservation equation gives the power output as

$$P = \rho \int \frac{1}{2}\mathbf{U} \cdot \mathbf{U}\mathbf{U} \cdot d\mathbf{A} \tag{2.11}$$

and it should be straightforward (after experience with the mass and momentum balances) to show that

$$P = \frac{1}{2}\rho Q_\infty \left(U_0^2 - U_\infty^2\right) \tag{2.12}$$

An alternative form of (2.12) can be found by applying Bernoulli's equation from the upstream face of the CV to a position just upwind of the blades to give

$$P_1 = \frac{1}{2}\rho\left(U_0^2 - U_1^2\right)$$

Similarly from just downwind of the blades to the far-wake, and noting that the velocity exiting the disk is the same as that entering;

$$P_2 = \frac{1}{2}\rho\left(U_\infty^2 - U_1^2\right)$$

(It is important to understand that Bernoulli's equation *cannot* be applied across the disk as the energy extracted from the air alters the Bernoulli constant on each streamline.) Using these two equations with (2.12) gives

$$P = Q_1(P_1 - P_2) = U_1(P_1 - P_2)\pi R^2 \tag{2.13}$$

Equations 2.6 and 2.13 can be combined to give

$$P = TU_1 \tag{2.14}$$

showing that the power is the product of the force on the disk and the air velocity through it. The correspondence between (2.14) and the relation between power, force, and velocity in engineering dynamics is obvious but it must be emphasised that (2.14) applies *only* to an ideal flow.

Combining (2.14) with (2.5a, b) and (2.12) gives the very interesting result that

$$U_1 = (U_0 + U_\infty)/2 \tag{2.15}$$

which is, because of the restriction on (2.14), applicable only to ideal flow. Equation 2.15 shows that half the expansion of the flow in terms of the velocity changes occurs before the blades and half in the wake, behind the blades.

2.7 Turbine Operating Parameters and Optimum Performance

The analysis of the previous section leads to equations for the turbine operating parameters that were introduced in Sect. 1.4, of which the most important is the power coefficient, C_P. From (2.3), (2.5a, b), (2.14) and (2.15):

$$C_P = \frac{P}{\frac{1}{2}\rho U_0^3 \pi R^2} = 4U_1^2(U_0 - U_1)/U_0^3 = 4a(1-a)^2 \qquad (2.16)$$

where a is the "axial induction factor" defined by

$$a = 1 - U_1/U_0 \qquad (2.17)$$

so that the larger the value of a the more deceleration occurs as the air goes through the blades. Maximum performance will occur when $\mathrm{d}C_P/\mathrm{d}a = 0$. From (2.16), this occurs when $a = 1/3$, and it immediately follows that

$$C_{P,\max} = 16/27 \approx 0.593, \quad \text{when } a = 1/3, U_1/U_0 = 2/3, \text{ and } U_\infty/U_0 = 1/3 \qquad (2.18)$$

for optimum performance. This is the Betz–Joukowsky limit. Its derivation shows that a turbine can never capture all the kinetic energy that would flow past the blade disk in the absence of the blades. All it can possibly do, according to (2.17), is to capture two-thirds of that wind (in terms of U_1/U_0), and convert eight-ninths of that into output power because $1 - (U_\infty/U_0)^2 = 8/9$. To do so, there must be significant expansion of the flow; the cross-sectional area of the far-wake when $a = 1/3$ is twice the blade disk area and three times the area of the wind captured by the blades. This expansion of an optimal wind turbine wake is large compared to the contraction that occurs in the wake of an efficient propeller, which is typically only 10%. Hovering rotors, which model helicopters in hover, have a wake contraction comparable to an optimum wind turbine's expansion. These comparisons are made to indicate that the operating range between the optimal λ and runaway, where wake expansion increases further with λ, is difficult to analyse. This issue is raised again in Chap. 7.

The derivation of the Betz–Joukowsky limit depends on major simplifications and assumptions about the air flow, principally in terms of steadiness, uniformity, and the neglect of viscosity, which cannot be strictly valid in practice. Nevertheless, no well-documented study of wind turbine power output has violated the limit in (2.18). From Sect. 1.4, the value of $C_{P,\max}$ for modern wind turbines is about 0.50. Furthermore, (2.18) is derived *without any reference* to the turbine blades themselves, so it is reasonable to state that the first job of a blade designer is to produce blades that result in an airflow as close as possible to the ideal. The next chapter considers the relationship between the flow over the blades, the forces acting on them, and a modification of the present analysis that is accurate at least up until the optimum performance point.

To finish this discussion of maximum performance, it is important to emphasise that C_P is, strictly, not an efficiency, so that the Betz–Joukowsky limit is *not* a limit on efficiency. Hopefully, the discussion in Sect. 1.1 of the amount of wind passing through the blades emphasises this point; improved performance only requires an increase in the amount of wind captured by the blades that outweighs any degradation in efficiency of conversion. Several methods have been proposed to increase flow capture, of which "diffuser augmentation" is probably the most promising.

A "diffuser augmented wind turbine" (DAWT) has a diffuser installed downstream of the blades (and sometimes a concentrator upstream) to force the wake to expand more rapidly than it would otherwise. This increases the air flow through the blades. It is easy to design a DAWT whose performance exceeds the Betz–Joukowsky limit. It is, however, generally thought that the structural and other problems of DAWTs, such as the difficulty in responding to changes in the wind direction, make them unsuitable in practice.

In a form similar to (2.16), the thrust coefficient, C_T, can be derived from Eqs. 2.5a, b and 2.11, giving

$$C_T = \frac{T}{\frac{1}{2}\rho U_0^2 \pi R^2} = 4a(1-a) \tag{2.19}$$

so that at optimum performance, $C_T = 8/9$, and has an upper limit of unity when $a = 1/2$ according to (2.19). In practice, it seems that C_T is about 0.9 at maximum performance, but there are a number of measurements of C_T exceeding unity as a increases above 1/3. Some of these are shown in Fig. 2.2. High thrust coincides with C_P decreasing towards the runaway value of zero, and increasing wake expansion. A further failing of the simple analysis is that Eq. 2.15 gives $U_\infty = 0$ when $a = 1/2$, and negative U_∞ for larger values, but, unfortunately there are no detailed measurements available for the high thrust region to suggest an appropriate modification to (2.15). The experimental data shown in the figure are the smoothed data from Table 7 of [2]. Additional data are given by Buhl [3] and Fig. 13.5 of [4]. The measurements must be treated with caution, because a was not measured directly, and the rotor swept area was a considerable fraction of the wind tunnel cross-sectional area, so that its blockage was high. The difference between the diamonds and the crosses in Fig. 2.2 is that the upstream "free-stream" velocity, equivalent to U_0, was used for the latter, but the measured "free-stream" velocity at the rotor was used for the former. Of course, the two would be equal in the absence of blockage. The consensus among the wind turbine

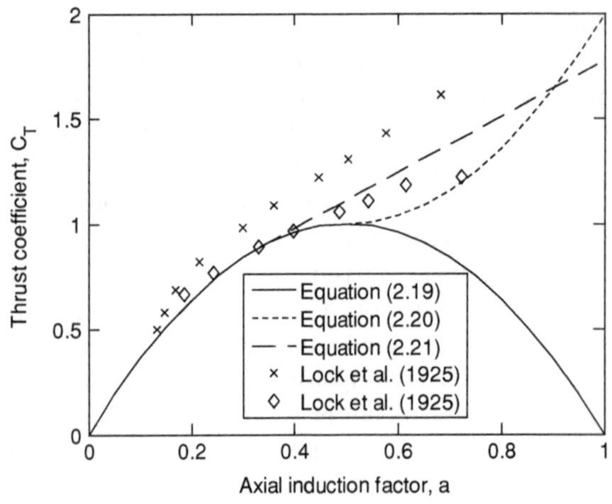

Fig. 2.2 Thrust coefficients at high axial induction

community is that the simple analysis leading to (2.16), the Betz–Joukowsky limit, and (2.18) is sufficiently accurate at least until the point of maximum power production. After that point, which is admittedly the designer's first goal, little is known about the details of the flow.

There have been a large number of suggested "corrections" to Eq. 2.19 advanced over the years, none of which are based on any proper physical consideration of the origin for the additional thrust or on the nature of the wake. One commonly used correction is due to Glauert; it consists of replacing (2.19) by

$$C_T = 4a(a-1) + 2 \tag{2.20}$$

whenever $a > 1/2$. As shown in Fig. 2.2, Eq. 2.20 is the mirror image of (2.19) about the line $C_T = 1$. Another "high thrust" equation is

$$C_T = \left\{ \begin{array}{ll} 4a(1-a) & a \leq a_c \\ 4a_c^2 + (1-2a_c)a & a > a_c \end{array} \right\} \tag{2.21}$$

where $a_c = 1/3$, which is also shown in Fig. 2.2. This equation was used by Clifton-Smith [5] and others and will be used for the blade element calculations in Chap. 5. Fortunately for performance analysis and design, the high thrust corrections are not critical. Another high thrust equation was proposed by Buhl [3] and used by Lanzafame and Messina [6].

The high thrust region deserves detailed examination because of the importance of the simple one-dimensional analysis for BET and the Betz–Joukowsky limit. It is possible that the basic wake structure of helical tip vortices trailing from the blades remains unaltered as the thrust increases, but that the internal structure of these vortices changes to alter the axial induction through the rotor and to absorb energy. Some evidence for this view was the finding of [7] that the near-wake of a model rotor at runaway had a significant velocity deficit, showing that kinetic energy had been extracted. However, the tip vortices also had sufficient angular momentum to absorb nearly all that energy and leave none to produce power. In other words, the simple assumption that the tip vortices are vortex lines whose structure is not important, which is appropriate for the one-dimensional flow at low thrust, quickly becomes incorrect. The importance of the wake structure in determining turbine performance is the main reason why the discussion on optimum performance in Chap. 5 starts with consideration of the far-wake.

2.7.1 Exercises

1. In deriving the equations of this chapter, it would have been much easier to use a CV whose outer face coincides with the bounding streamtube. Can you think of reasons why this CV was not used?
2. List the assumptions made in deriving the equations of this chapter. Which ones do you consider important and possibly responsible for the breakdown of the analysis at high wake expansion?

3. If the pressure in the far-wake was not ambient (or zero gauge pressure), how would Eq. 2.4 be altered?
4. At the end of Sect. 2.2 it was stated that any circumferential velocity behind the blades does not appear in the conservation mass equations. Explain why this is so.
5. If there is swirl in the far-wake, i.e. non-zero W_∞, then the pressure in the far-wake, P_∞, is modified according to

$$\frac{dP_\infty}{dr} = \frac{\rho W_\infty^2}{r}$$

where r is the radius. The derivation of this "centrifugal force" equation can be found in any standard fluid mechanics text, e.g. [1]. Under what conditions is Eq. 2.12 unaffected by swirl?
6. What effect does swirl have on the thrust equation, (2.5a, b)?
7. Why did this chapter contain no discussion of Reynolds number, torque coefficient and tip speed ratio?
8. An inventor claims to have developed a water turbine of 1 m diameter that produces 5 kW in a tidal flow of five knots (1 knot = 0.515 m/s). The density of water is 1000 kg/m^3. Do you believe the inventor's claim?
9. Starting from Eq. 2.14, derive the relation between C_T and C_P for ideal flow.
10. It is generally held that the results such as shown in Fig. 2.2, imply that the thrust coefficient C_T never exceeds a value of 2. Using this constraint, estimate the maximum horizontal force exerted on the tower by the 5 m diameter turbine described in Chap. 1 for wind speeds of 10 and 25 m/s.
11. Is there an optimum value of C_Q as there is for C_P?

References

1. White F (2011) Fluid mechanics, 7th edn. McGraw Hill, New York
2. Lock CNH, Bateman H, Townend HCH (1925) An extension of the vortex theory of airscrews with applications to airscrews of small pitch, including experimental results. ARC reports & memoranda no 1014
3. Buhl ML Jr (2005) A new empirical relationship between thrust coefficient and induction factor for the turbulent wind state, NREL/TP-500-36834, National Renewable Energy Laboratory, Golden, http://www.nrel.gov/docs/fy05osti/36834.pdf. (accessed 1 Oct 2010)
4. Leishman JG (2006) Principles of helicopter aerodynamics, 2nd edn. Cambridge University Press, Cambridge
5. Clifton-Smith M (2009) Wind turbine blade optimisation with tip loss correction. Wind Eng 33:477–496
6. Lanzafame M, Messina M (2007) Fluid dynamics wind turbine design: critical analysis, optimization and application of BEM theory. Renew Energy 32:2291–2305
7. Ebert PR, Wood DH (2002) The near wake of a model horizontal-axis wind turbine at runaway. Renew Energy 25:41–54

Chapter 3
Blade Element Theory for Wind Turbines

3.1 Introduction

The conservation equations of fluid mechanics were used in the last chapter to derive equations for the power output and thrust for a turbine modeled as an actuator disk. When it is assumed that there is no radial dependence in the flow or over the disk, these equations lead to the well-known Betz–Joukowsky limit on wind turbine performance. However, they do not consider the forces acting on the individual blades which give rise to the thrust and torque, and hence power. These are obviously the main quantities required for analysis of wind turbine performance and strength and are necessary for aerodynamic and structural design.

The traditional way to extend the analysis of Chap. 2 is to divide the flow through the blades into a number of concentric annular streamtubes. In practice, the number of streamtubes should be large enough to provide a good approximation to the variations in velocity, chord, and twist along a blade. Experience shows that typical performance analyses can be done accurately with between 10 and 20 blade elements. The conservation equations for these streamtubes are easily-recognised generalisations of those derived in Chap. 2 for mass, axial and angular momentum, and energy. The velocity and pressure within each streamtube do not vary with radius but may vary from one streamtube to the next. Each streamtube intersects a blade element, generating a lift and drag. The next section makes explicit a number of the fundamental assumptions of blade element theory, of which the most important is that blade elements behave as aerofoils. It is, therefore, necessary to review the important aspects of aerofoil theory with special reference to small wind turbines; this is done in Chap. 4. Section 3.3 develops the important conservation equations in annular streamtube form, starting from the analysis of Chap. 2. Section 3.4 considers the forces acting on the blade elements. The implementation of these equations in a Matlab computer program to predict wind turbine performance is covered in Sect. 3.5. Section 3.6 describes this program which allows the use of any number of blade elements. The M-files of all listed programs are available in the online materials (http://extras.springer.com).

D. Wood, *Small Wind Turbines*, Green Energy and Technology,
DOI: 10.1007/978-1-84996-175-2_3,

This chapter is a brief introduction to the theory. A much more extensive explanation is given in Chap. 3 of [1]. For this chapter there are no significant features that are specific to small turbines.

3.2 Some Assumptions of Blade Element Theory

To extend the analysis of Chap. 2, use is made of the following assumptions:

- The flow in each streamtube is independent of that in other streamtubes, and
- The forces acting on each blade element are the same as those on an aerofoil of the same section, angle of attack, and effective velocity.

It is easy to demonstrate that both these assumptions can be in error. The first assumption is violated if there is a radial variation in the velocity through the blades, as this will cause a radial pressure gradient and each streamtube will therefore exert a force on its two neighbours. Fortunately this force is redistributive, i.e. it sums to zero over the whole flow, and so ignoring it should not cause too severe an error.

The second assumption is much more intriguing. As will be made clear in Chap. 4, an aerofoil is a two-dimensional body in an infinite flow that is uniform away from the region influenced by the body. Such a situation can never occur in a wind turbine because the blades are always separated by a finite distance in the azimuthal direction. The measure of the importance of this effect is the solidity which is defined in Sect. 3.5. Another difference from aerofoil behaviour is that the flow over blade elements can remain attached at angles of attack that would cause an aerofoil to stall, see Sect. 3.12 of [1]. This so-called "stall delay" is an empirical fact, but its cause is not clear. It is usually argued that the Coriolis and centrifugal forces in the boundary layers on the rotating blades are responsible. However, the most common correction of aerofoil lift and drag for stall delay, see Eq. 3.190 of [1], involves only the solidity. In other words, stall delay supposedly scales on a parameter that does *not* measure the centrifugal and Coriolis forces because it does not contain Ω. Furthermore, stall delay predominantly occurs towards the hub, so it is likely that solidity, which is usually larger near the hub and also delays separation, is at least partially responsible. Fortunately, stall delay does not significantly influence optimum turbine performance, so no attempt will be made to include its effects in the blade element calculations.

3.3 The Conservation Equations for Annular Streamtubes

These equations can be obtained by applying the vector equations in the last chapter to the annular streamtube at radius r, and radial extent dr, as shown in Fig. 3.1.

The blade tip radius is R. The streamtube thickness is dr_0 upstream, dr at the blades, and dr_∞ in the far-wake. Note that the streamtube is annular and it is

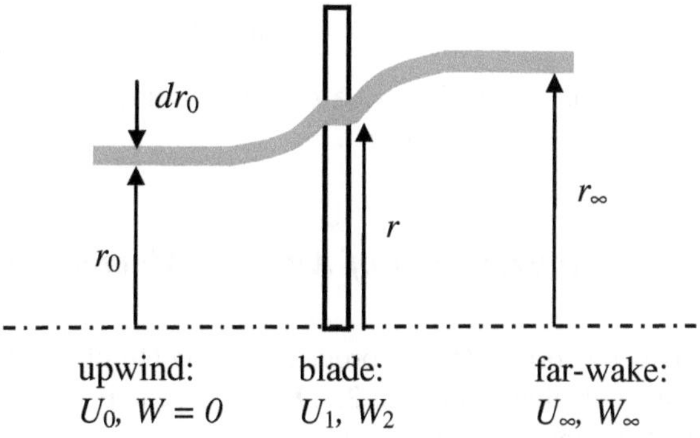

Fig. 3.1 Annular streamtube intersecting a blade element

assumed that $dr \ll r$ typically. Any velocity in the radial direction is ignored, but the circumferential or swirl velocity will be included in the analysis. The conservation equations for mass, momentum, angular momentum, and energy are now considered in turn.

3.3.1 Conservation of Mass

Dividing the conservation of mass Eq. 2.1 by the density and applying it to the streamtube whose flow area is approximately $2\pi r dr$, gives

$$U_0 2\pi r_0\, dr_0 = U_1 2\pi r\, dr = U_\infty 2\pi r_\infty\, dr_\infty$$

or, in a form analogous to (2.3),

$$U_0 r_0\, dr_0 = U_1 r\, dr = U_\infty r_\infty\, dr_\infty \tag{3.1}$$

3.3.2 Conservation of Momentum

Because the force of main interest is in the direction of the wind and the turbine's axis, it is easier to revert to scalars. Using Eq. 3.1, the contribution to the axial thrust, T, from the streamtube is:

$$dT = \rho U_0 U_0 2\pi r_0\, dr_0 - \rho U_\infty U_\infty 2\pi r_\infty\, dr_\infty = 2\pi\rho U_1 r(U_0 - U_\infty)\, dr$$

Note that this is the total force acting on the N blade elements that intersect this streamtube. Using Eqs. 2.15 and 2.17, the equation can be rewritten as

$$dT = 4\pi r\rho U_0^2 a(1 - a)\, dr \tag{3.2}$$

a is the axial interference factor, sometimes called the "axial inflow factor", defined in Sect. 2.6:

$$a = 1 - U_1/U_0 \tag{3.3}$$

so that the larger the value of a the greater the reduction in speed as the wind passes through the blades.

3.3.3 Conservation of Angular Momentum

In vector form, $d\boldsymbol{Q}$, the torque acting on the blade elements within the streamtube can be obtained from Eq. 2.7. From this can be deduced the scalar equation for the contribution to Q, the torque acting about the axis of rotation:

$$dQ = \rho r_\infty W_\infty U_\infty 2\pi r_\infty \, dr_\infty \tag{3.4}$$

assuming that there is no swirl upstream of the blades. (This is an important assumption that will be examined in more detail in Chap. 4.) Downstream of the blades, the angular momentum of the streamtube is conserved so $rW_2 = r_\infty W_\infty$. Using this relationship and conservation of mass

$$dQ = 2\pi\rho U_0(1-a)W_2 r^2 \, dr = 4\pi\rho U_0(1-a)a'\Omega r^3 \, dr \tag{3.5}$$

where $W_2 = 2a'\Omega r$ defines (twice) the rotational interference factor. The geometric significance of a and a' are discussed in the next section. Note that the average W seen by the blades is

$$W = (W_0 + W_2)/2 = W_2/2 = a'\Omega r \tag{3.6}$$

where a' is the rotational interference factor.

3.4 The Forces Acting on a Blade Element

The analysis of the previous two sections gives the velocity components for each blade element at radius r. The situation is summarised in Fig. 3.2. The velocity in the wind direction is U_1 and the circumferential velocity is the sum of Ωr and W as defined in Eq. 3.6. Adding these velocities vectorially, and ignoring any radial velocity, gives the non-dimensional velocity U_T:

$$U_T^2 = (1-a)^2 + [(1+a')\lambda_r]^2 \tag{3.7a}$$

where λ_r is the local speed ratio (of the blade element)

$$\lambda_r = r\Omega/U_0 = \lambda r/R \tag{3.7b}$$

U_T is usually called the "total" or "effective" velocity as seen by the blade element. α is the angle of attack, which is sometimes called the angle of incidence. This is one of the three important angles defined in Fig. 3.2: θ_P the twist, is the angle between the plane of rotation of the blade and the element's chord line. Sometimes θ_P is termed the pitch angle, but, in this text, "pitch" will signify a constant, global change in θ_P caused by alteration of the blade's attachment at the hub. Finally, ϕ is the inflow angle between U_T and the plane of rotation. From the geometry

$$\theta_P + \alpha = \phi \tag{3.8}$$

Note very carefully that Fig. 3.2 does NOT indicate the location of the effective velocity relative to the blade element or the line of action of the forces. In aeronautical applications this line of action can be very important, for example in determining the longitudinal stability of an aircraft. For wind turbines, however, its location has much less significance.

Figure 3.3 shows the resulting lift and drag. By definition, the lift acts at right angles to U_T and the drag acts in the direction of U_T. Chapter 4 shows that the magnitude of the lift is many times that of the drag for well-designed aerofoils. Since the primary purpose of the forces on the blade element is to produce a torque about the axis of rotation, or equivalently, a circumferential force in the direction of rotation, the figure indicates the necessity of maximising the lift and minimising the drag. Very simply, drag acts to *reduce* the torque produced by the lift: the key to wind turbine performance is the *ratio* of the lift to drag, rather than the individual values.

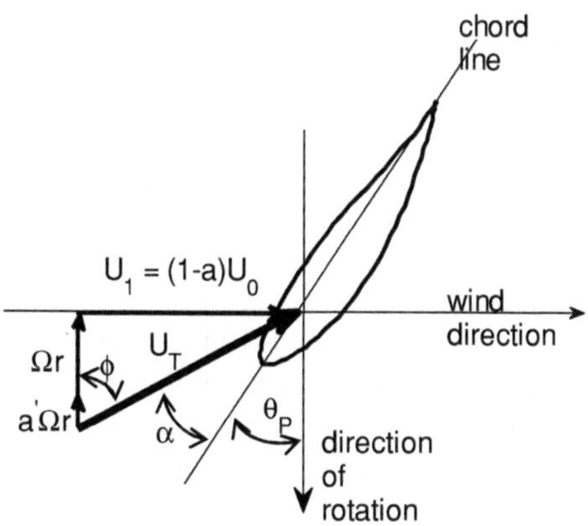

Fig. 3.2 Velocities for blade element at radius ***r***

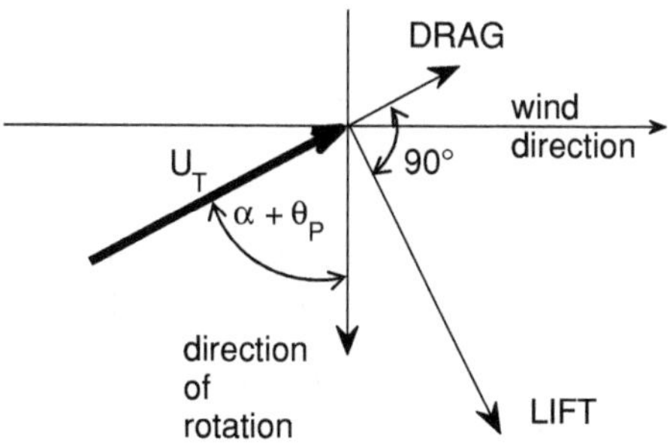

Fig. 3.3 Lift and drag on a blade element

Because wind turbines operate at high values of λ, typically in the range of 7–10, Ωr at the tip is about ten times greater than U_1. At the hub, Ωr is nearly zero, so that θ_P must vary significantly with radius to maintain the angle of attack, α, at reasonable values to avoid flow separation. Chapter 4 shows that efficient operation of wind turbines occurs over a limited range of α.

The basic assumption is that the lift and drag acting on the blade element are the same as those on an aerofoil of the same section, angle of attack, and effective velocity. From the definitions of the lift and drag coefficients, C_l and C_d respectively:

$$\text{LIFT} = \frac{1}{2}\rho U_T^2 C_l c \quad \text{and} \quad \text{DRAG} = \frac{1}{2}\rho U_T^2 C_d c \tag{3.9}$$

where c is the chord; its precise definition will be given in Chap. 4 along with further information on the coefficients. It is next necessary to resolve the lift and drag into the circumferential and axial components of interest to the wind turbine designer. For an N-bladed turbine, the total thrust on N blade elements is

$$dT = \frac{1}{2}\rho U_T^2 cN(C_l \cos\phi + C_d \sin\phi)\, dr = \frac{1}{2}\rho U_T^2 cNC_a\, dr \tag{3.10}$$

where $C_a = C_l \cos\phi + C_d \sin\phi$ and the torque due to the circumferential force is

$$dQ = \frac{1}{2}\rho U_T^2 cN(C_l \sin\phi - C_d \cos\phi) r dr = \frac{1}{2}\rho U_T^2 cNC_{a'} r dr \tag{3.11}$$

where $C_{a'} = C_l \sin\phi - C_d \cos\phi$.

Equations 3.10 and 3.11 are the basic blade element equations. A number of modifications to them have been proposed in the century following their original development for propeller analysis; some of these will be considered in Chap. 5 following the discussion of aerofoil lift and drag in Chap. 4.

Example 3.1 Suppose a turbine is operating at its maximum efficiency where $a = 1/3$ and a' is negligible. If $\lambda = 7$, estimate the θ_P necessary to achieve $\alpha = 6°$ at $r/R = 0.25$ and $r/R = 1$.

Answer for r/R = 0.25 In the blade element diagram of Fig. 3.2, $U_1/U_0 = 2/3$ and $\lambda_r = r\Omega/U_0 = 1.75$. Now $\phi = \tan^{-1}(2/(3\lambda_r)) = 20.85°$, so that, from Eq. 3.7a, b, $\theta_P = 14.85°$.

Answer for r/R = 1.0 In the blade element diagram of Fig. 3.2, $U_1/U_0 = 2/3$, $\lambda = R\Omega/U_0 = 7.0$, so that $\phi = 5.44°$. From Eq. 3.7a and 3.7b, $\theta_P = -0.56°$.

These answers show the necessity for a significant variation in twist along a well-designed wind turbine blade.

3.5 Combining the Equations for the Streamtube and the Blade Element

Equations 3.2 and 3.5 for the wake must be solved with (3.10) and (3.11) for the blade element. One way of doing this is through the following relationship for ϕ:

$$\tan\phi = \frac{U_1}{\Omega r(1+a')} = \frac{1-a}{\lambda_r(1+a')} \tag{3.12}$$

which follows from the velocity triangle Fig. 3.2. λ_r is the local speed ratio from Eq. 3.7b which obviously varies from zero at the axis of rotation to λ at the blade tip. If values are assumed for a and a', ϕ can be found from (3.12) and then C_l and C_d determined and so on. This is an iterative process because (3.2) and (3.10) can be combined to give a_n, the new estimate for a for the nth iteration, as

$$a_n(1-a_n) = f_a = U_T^2 C_a \sigma / \left(4U_0^2\right) \tag{3.13}$$

where the quantities on the right hand side of (3.13) are those from the $(n-1)$ iteration. In Eq. 3.13, σ is defined as

$$\sigma = Nc/(2\pi r) \tag{3.14}$$

which is the local solidity mentioned in Sect. 3.2. f_a in (3.13) plays the same role for the blade element as does C_T for the whole turbine in the analysis of Chap. 2. Thus, for example, it is easy to modify (3.13) for the high thrust region in the same way that Eqs. 2.20 and 2.21 can modify (2.19).

Combining (3.5) and (3.11) gives

$$a' = aC_{a'}/(C_a\lambda_r) \tag{3.15}$$

This allows the iteration to be solely based on a until the momentum and angular momentum fluxes in the flow over the blades balance the thrust and torque, respectively, on a blade element.

3.6 Matlab Programs for Blade Element Analysis

This section describes and lists three simple Matlab programs that implement the equations derived in the last two chapters. Use of the programs will be postponed until aerofoil properties are more extensively discussed in Chap. 4, including the circulation, and their consequences. The programs are available from the online materials site (http://extras.springer.com).

All lengths are normalised by the tip radius and all velocities by the wind speed, unless specifically stated otherwise in the definition of the appropriate variable in the comments at the start of each program.

Table 3.1 Blade geometry from Anderson et al. [2]

Radius (cm)	Chord (cm)	Twist (°)
13.33	25.02	24.21
23.25	23.18	21.06
29.75	19.98	15.87
36.25	17.32	11.92
42.75	15.12	8.96
49.25	13.33	6.77
55.75	11.86	5.18
62.25	10.67	4.02
68.75	9.71	3.16
75.25	8.92	2.49
81.75	8.26	1.93
88.25	7.71	1.44
94.75	7.23	0.99
101.25	6.80	0.59
107.75	6.41	0.25
114.25	6.04	−0.06
120.75	5.70	−0.36
127.25	5.38	−0.67
133.75	5.08	−0.98
140.25	4.84	−1.28
146.75	4.66	−1.59
150.0	4.60	−1.74

The main "program", simple_power_calc.m has the single input argument of the number of blades (Numb). All blade elements have the same radial width, dr. tcdist.m determines the chord and twist distributions for any number of blade elements. The particular polynomial equations for the chord and twist are a fit to the data for a two-bladed 3 m diameter turbine that was extensively field- and wind-tunnel tested by Anderson et al. [2], see the listings in Table 3.1. The measurements of the turbine's performance will be discussed in detail in Chap. 5. Note that tcdist.m allows the user to specify a pitch angle, which is a global, constant alteration to the twist.

Having read in the blade geometry, simple_power_calc.m begins an outer loop to allow the user to vary the windspeed (U0) and then the tip speed ratio (lambda) in the inner loop. The lift and drag coefficients from the subroutine in LandD_0012.m which contains data fits for the NACA 0012 profile are discussed in the next chapter. These fits are restricted both in angle of attack (aoa) and Reynolds number (Re) and *should never be used outside their range of validity*.

The iteration is extremely simple: the new and old values of a (a) are simply averaged to find the appropriate values for the next iteration. Convergence in a is determined using the tolerance, tol.

The listing of simple_power_calc.m follows:

```
function simple_power_calc(Numb)

% Program to implement blade element/one-dimensional wake
%analysis for a horizontal-axis wind turbine with any number of
%blades of any length. Unless otherwise specified, all lengths are
%normalised by the tip radius (r_tip) and all velocities by the wind
%speed, U0.  In this version the blade element width (delr) is equal
%for all  blade elements.

% Function argument is:
%       Numb - number of blades

% Main variables are:
%  a - axial interference factor
%  adash - rotational interference factor
%  aoa - angle of attack of blade element
%  chord - chord of blade element
%  C_a - factor for elemental thrust
%  C_adash - factor for elemental torque
%  Cd - drag coefficient
%  Cl - lift coefficient
%  Cp - power coefficient
%  Ct - thrust coefficient
%  delr - radial width of blade element
%  delthr - thrust on blade element
%  deltor - torque on blade element
%  lambda - tip speed ratio (TSR)
%  lamr - local speed ratio of blade element, Eq. (3.7b)
%  gam - circulation of blade element
%  nbes - number of blade elements
%  phi - inflow angle between Ut and plane of rotation
%  Re - blade element Reynolds number
%  r_hub - radius of hub
%  r_tip - radius of tip (m)
%  rad - radius of midpoint of blade element
%  sigma - local solidity, Eq. (3.14)
%  twist - angle between chord line and plane of rotation
%  Ut - effective velocity at blade element, Eq. (3.7a)
%  U0 - wind speed (m/s)
visc = 1.5e-5; % Kinematic viscosity of air (m^2/s)
rho = 1.2;    % Density of air (kg/m^3)
tol = 1.e-4;  % Convergence tolerance for BE analysis
in = load('rad_ch_tw.dat'); % data file with radius, chord, and twist
rad=in(:,1);    % Recover the radius of the blade elements
nbes = rad(1);  % First entry is nbes
```

```
rad(1) = [];     % Remove from array for radius
chord=in(:,2);  % recover the chord of the blade elements
r_tip= chord(1); % First entry is r_tip
chord(1) = [];
twist=in(:,3);  % Recover the twist of the blade elements
r_hub = twist(1); % First entry is r_hub
twist(1) = [];
delr=rad(2)-rad(1); % Determine width of blade elements

out_format=' %7.4f %3d  %7.2f  %7.3f  %7.3f  %8.5f'... ' %8.5f  %8.3e \n';

U0 = 100.0;
while U0 > 0.0
  U0 = input(' Enter the wind speed in m/s: end with -ve: ');
  if U0 < 0.0, break; end
     while U0 > 0.0
        lambda = input(' Enter TSR: end with -ve: ');
        if lambda < 0.0, break; end
        thrust = 0.0; torque = 0.0;
        fprintf('\n')
        fprintf('  Radius  iter.  aoa      a      Cl       Cd')
        fprintf('      deltor     Re \n')
        a = 0.3; % Initialise a
        for i = 1: nbes % Loop over each blade element
           adash = 0.0; deltor = 0.0;
           lamr = lambda*rad(i);  % Local speed ratio, Eq. (3.7b)
           sigma = 0.5*Numb*chord(i)/pi/rad(i);  % Local solidity
           diffa = 200*tol*a;

           while diffa > tol*a
              phi = atan((1 - a)/(1 + adash)/lamr) ; Eq.(3.12)
              cosphi = cos(phi);
              sinphi = sin(phi);
              aoa = phi*180.0/pi - twist(i);Eq.(3.8)
              Ut = sqrt((1-a)*(1-a) + (lamr*(1 + adash))^2);
              Re = Ut*U0*chord(i)*r_tip/visc ; % Reynolds number
              [Cl, Cd] = LandD_0012(aoa, Re); % Find Cl and Cd
              C_a = Cl*cosphi + Cd*sinphi; % For axial force
              C_adash = Cl*sinphi - Cd*cosphi;  % tangential force
% Balance axial momentum and blade thrust, Eq. (3.13), to find
% new a
              faca = Ut*Ut*sigma*C_a;
% Use Glauertès empirical correction when a > 0.5, Eq. (2.20)
              if faca > 1.0
```

```
                newa = (1 + sqrt( faca - 1))/2;
            else
                newa = (1 - sqrt(1 - faca))/2;
            end
            diffa = abs(a - newa);
            a = 0.5*(a + newa); % Average old and new a
            adash = a*C_adash/(C_a*lamr); % Eq. (3.15)
        end
        Ngam = 0.5*Numb*Ut*Ut*chord(i)*C_adash/(1 - a);
        delthr = Numb*Ut*Ut*chord(i)*delr/pi; % Eq. 3.10
        deltor = delthr*rad(i)*C_adash; % Eq. 3.11
        delthr = delthr*C_a; % Complete Eq. 3.10
        thrust = thrust + delthr; % Sum the rotor thrust
        torque = torque + deltor; % Sum the rotor torque
        fprintf(out_format, rad(i), j, aoa, a, Cl, Cd, deltor, Re)
      end
      cp = torque*lambda; % Find the power coefficient
      fprintf(' \n')
      fprintf(' Cp = %5f, Ct = %5f \n', cp, thrust)
      power = 0.5*cp*rho*U0^3*pi*r_tip^2; % Find Power in Watts
      thrust = 0.5*thrust*rho*U0^2*pi*r_tip^2; % Find Thrust in N
      fprintf(' Power = %5e Watts, Thrust = %5e Newtons \n',...
        power, thrust)
      fprintf(' \n')
   end
end
```

This program requires the input data file "rad_ch_tw.dat" which contains the radius, chord, and twist of the blade elements. To simulate the experiments of Anderson et al. [2], tcdist.m uses a curve fit to the data in Table 3.1:

```
function tcdist(nbes, pitch, r_tip, r_hub_in)
% M -file to give chord and twist distribution of the blade used by
% M. B. Anderson et al. (1982).
% The output is written to rad_ch_tw.dat where r_tip is in m
% radius is normalized by tip radius and angles are in degrees.
% Variables are:
% nbes - number of blade elements
% pitch - blade pitch or setting angle in degrees
% r_tip - radius of tip (m)
% r_hub_in - radius of hub (m)
% rad(i) - radius of the centre of element i
```

```
% chord(i) - chord of blade element i
% twist(i) - twist of blade element i

r_hub = r_hub_in/r_tip;
delr = ( 1.0 - r_hub)/nbes;
for i = 1:nbes
        if (i == 1)
                rad(1) = r_hub + delr/2;
        else
                rad(i) = rad(i-1) + delr;
        end
        chord(i) = chd(rad(i));
        twist(i) = t1(rad(i)) + pitch;
end
data_out =[rad' chord' twist'];
% Make the first line of the data contain nbes r_tip and r_hub
data_out =[nbes r_tip r_hub_in;data_out];
save('rad_ch_tw.dat','data_out','-ascii')

function out = t1(x)
% Twist distribution. Fit is 4th order poly to r/R = 0.7, then linear.
if (x <= 0.7)
        out = 54.16632 - x*(307.42939 -x*(719.549614 - ...
                x*(785.971096-x*326.673372)));
else
        out = 5.318999 - 7.059999*x;
end
return

function out = chd(x)
% Chord distribution - 4th order poly fit to data in Anderson et al.
out = 5*(0.16165732 - x*(0.5847727 - x*(1.0327255 –...
                x*(0.8756711 -x*0.2844545))))/3;
return
```

The running and output of tcdist.m is shown in the box below. Note that the double greater than signs ">>" is the Matlab command prompt:

```
>> tcdist(15,0,1.5,0.15)
>> type rad_ch_tw.dat

1.5000000e+001 1.5000000e+000 1.5000000e-001
1.3000000e-001 1.6874554e-001 2.4727410e+001
1.9000000e-001 1.3699396e-001 1.6765225e+001
2.5000000e-001 1.1239713e-001 1.1276093e+001
3.1000000e-001 9.3604305e-002 7.6139593e+000
3.7000000e-001 7.9412182e-002 5.2343789e+000
4.3000000e-001 6.8764925e-002 3.6945145e+000
4.9000000e-001 6.0754160e-002 2.6531376e+000
5.5000000e-001 5.4618973e-002 1.8706279e+000
6.1000000e-001 4.9745912e-002 1.2089738e+000
6.7000000e-001 4.5668986e-002 6.3177214e-001
7.3000000e-001 4.2069663e-002 1.6519973e-001
7.9000000e-001 3.8776877e-002 -2.5840021e-001
8.5000000e-001 3.5767019e-002 -6.8200015e-001
9.1000000e-001 3.3163942e-002 -1.1056001e+000
9.7000000e-001 3.1238963e-002 -1.5292000e+000
```

Finally, the lift and drag coefficients are required as a function of the Reynolds number and angle of attack. A very simple function to do this is LandD_0012.m which uses the data correlations for the NACA 0012 aerofoil based on those of [3] as described in Chap. 4. It will be seen in Chap. 5 that the inputting of measured lift and drag is more cumbersome than the simple situation shown here.

```
function [cl, cd] = LandD_0012(aoa, Re)
% Function to calculate the lift, cl, and drag, cd, coefficient of
% a NACA0012 aerofoil using the correlations of McCroskey (1987).
  if aoa > 12.0, aoa = 12.0; end % Eq. 4.3 for Cl is only valid...
  if aoa < -12.0 aoa = -12.0; end% for aoa < |12 deg|.
  cl = aoa*(0.1025 + 0.00485*log10(Re/10^6));
  cd0 = 0.0044 + 0.018*Re^(-0.15); % Eq. 4.4 for minimum Cd
  delcd = (cl/1.2)^2*0.009; % A data fit to obtain Cd at other angles
  cd = cd0 + delcd;
return
```

The more complex and realistic power_calc.m and its associated programs are presented and described in Chap. 5. They have the same basic layout as shown here. In other words, the blade geometry and the aerofoil lift and drag data are determined separately, to allow the same main program to handle a wide range of different configurations and aerofoils. It is common for large wind turbines for the aerofoil section to change along the blade, but this does not occur often for small turbines. The reasons for this difference are explored in Chaps. 4–6, but for now it is sufficient to note the simplicity of using the same section in the blade element program.

3.7 Some Consequences of the Blade Element Equations

In preparation for the next chapter on Reynolds number effects, it is instructive to look at Eq. 3.11 for maximum torque, given N and Ω. Ignoring the drag, the product $U_T c C_l$ must be maximised. This product is proportional to ReC_l which, therefore, should remain constant as the wind speed varies. ReC_l is sometimes called the reduced Reynolds number, and indicates that an aerofoil section need not have the highest possible C_l; in fact, a smaller C_l can lead to larger Re and hence to a smaller performance penalty as the wind speed decreases.

Chapter 5 describes the implementation of the programs developed in this chapter; the preceding paragraph indicates the need to consider aerofoil behaviour before proceeding. However, this is a good opportunity to consider briefly the chord and twist distributions in Table 3.1. The reason for the decrease in twist as radius increases was shown by example in Sect. 3.4: it is the need to keep each blade element at an efficient angle of attack while U_T turns from the axial towards the tangential direction with increasing radius. An "efficient" angle of attack is likely to be close to the optimum angle that maximises the ratio of lift to drag. The reason for the approximately inverse relation between chord and radius will become clearer after considering the blade's circulation at the end of Chap. 4.

3.7.1 Exercises

1. Make a list of the assumptions of blade element theory as presented in this chapter.
2. What are the units of dT in Eq. 3.2 and dQ in (3.5)?
3. A turbine is operating at $\lambda = 9$ in a wind of 10 m/s. Estimate the minimum U_T (near the hub at, say, $0.1R$) and maximum U_T at the tip.
4. How would Fig. 3.2 change if a propeller, rather than a wind turbine, were being analysed?
5. Redo the calculations of Example 3.1 for $\lambda = 3$ and 10. What do the results suggests for a strategy to control the power output?
6. Interpret the C_P versus λ curve (Fig. 1.6) for the Vestas V80 in terms of the variation in typical angle of attack of a blade element near the blade tip. Assume the pitch angle does not change.
7. The power curve for the V80 in Fig. 1.5 is for constant Ω operation. Determine the qualitative behaviour of α as U_0 increases.
8. What are the limitations on the angle of attack for the equations in LandD_0012.m?
9. Derive the equation for $N\Gamma$ used in simple_power_calc.m:

```
Ngam = 0.5*Numb*Ut*Ut*chord(i)*C_adash/(1 - a);
```

References

1. Burton T, Sharpe D, Jenkins N, Bossanyi E (2001) Wind energy handbook. Wiley, New York
2. Anderson MB, Milborrow DJ, Ross JN (1982) Performance and wake measurements on a 3 m diameter horizontal axis wind turbine. In: 4th International Syrupon Wind Energy Systems, pp 113–135
3. McCroskey W (1987) A critical assessment of wind tunnel results for the NACA 0012 airfoil. NASA Technical Memorandum 100019

Chapter 4
Aerofoils: Lift, Drag, and Circulation

4.1 Introduction

The analysis in Chap. 3 assumed that blade elements behave as aerofoils. Thus wind turbine thrust and power depend upon the lift and drag coefficients, C_l and C_d respectively, of the aerofoil sections that comprise each blade. For a great many aerofoils, these coefficients are known from wind tunnel investigations, and, at least in principle, can be used immediately for power and thrust calculations. Although computational analysis is routine for aerofoils, the author's opinion is that computed C_l and C_d results are not to be trusted, particularly at the low Reynolds numbers that characterise small wind turbines. Some experimental data are also not to be trusted because of poor quality equipment or test procedures, e.g. [1]. It is important to attempt to establish the veracity of any aerofoil data used for analysis or design.

This chapter considers the basic features of aerofoils and emphasises aspects of their performance that are significant for small wind turbine application. These are behaviour at high angles of attack and low Reynolds number, *Re*, as well as the relation between lift and circulation. A very good introduction to low *Re* aerodynamics at low angles is Chap. 2 of [2].

It is important to remember that an aerofoil is two-dimensional and that, as indicated in Chap. 3, there are always assumptions involved in applying two-dimensional data to rotating, complex three-dimensional geometries such as wind turbine blades. Some of these assumptions will be considered in the next chapter.

4.2 Geometry and Definition of Aerofoils

Figure 4.1 shows four members of the NACA "four digit" series. NACA—the U.S. National Advisory Committee on Aeronautics—was the forerunner to NASA, and was very active in aerofoil development after the First World War. The four

D. Wood, *Small Wind Turbines*, Green Energy and Technology,
DOI: 10.1007/978-1-84996-175-2_4,

digit family of aerofoils was developed in the late 1920s. It has the advantage of being analytically described, whereas most modern sections are specified by their surface co-ordinates. The four digit family illustrates the basic terminology of aerofoils which arose from the manually-intensive methods of wing and propeller blade layout used in the early days of the aircraft industry.

The extreme forward point on the aerofoil is the leading edge, which is on the left for the four aerofoils in Fig. 4.1; the airflow is from left to right and the lift, l, acts upwards. Virtually all aerofoils have a sharp trailing edge, at the right hand end of the sections in Fig. 4.1. The straight line joining the leading and trailing edges is the chord line of length, c. The angle of attack, α, which was introduced in the previous chapter, is measured between the free stream velocity, U_0, and the chord line. If U_0 were in the horizontal direction, a positive α would occur by raising the leading edges. The chord line was normally the first quantity to be laid out for an aircraft wing. Then the mean line, shown dashed in Fig. 4.1, was added. It lies mid-way (the mean position) between the upper and lower surfaces of an aerofoil. The maximum distance between the chord line and the mean line is called the camber. Finally, the aerofoil thickness is added. The NACA four digit aerofoils are no longer used extensively, but the 0012 is probably the most studied aerofoil in history, see [1]. As indicated by the coincidence of the mean line and chord, this aerofoil is symmetric, so that it produces no lift at zero angle of attack. This is a very useful property for aircraft tail fins, as well as helicopter and other blades. The "12" in the designation indicates that this aerofoil, as well as the other three shown, has a maximum thickness, t, of 12% of the chord, although some sections specifically designed for the root areas of wind turbine blades, are thicker to accommodate the large centrifugal and other stresses in this region.

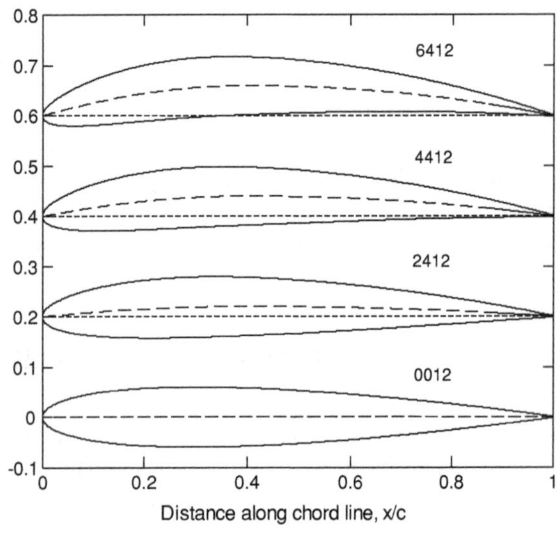

Fig. 4.1 Four examples of the NACA 4-digit series. The *solid line* denotes the surface, the *straight line* is the chord line and the *dashed line* is the mean line. All aerofoils have the same thickness (12%), but camber increases upwards. Each aerofoil is displaced upwards by 0.2 units

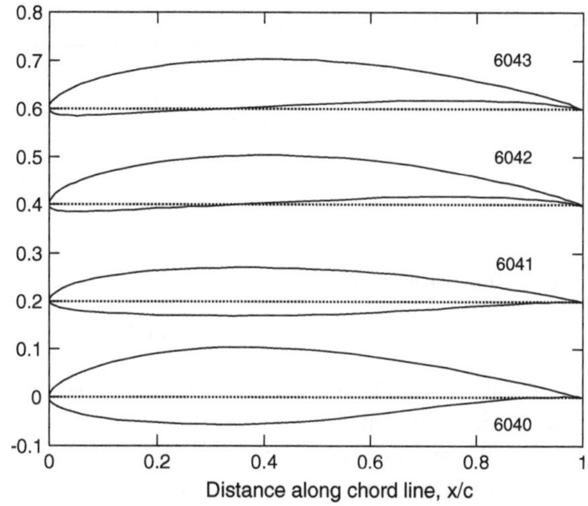

Fig. 4.2 The SG family of aerofoils for small wind turbines. The *straight line* is the chord line. The camber and thickness are in Table 4.1

The first of the four numbers indicates the camber as a percentage of c; this is zero for a symmetric section. The last aerofoil in Fig. 4.1, the 6412, has 6% camber, which is about the upper limit for practical aerofoils. The second number in the aerofoil designation, 4 in every case, indicates that the camber occurs at 40% of c. This is the position of maximum thickness for all members for the NACA four digit family.

Figure 4.2 shows the more modern SG aerofoils designed by Professor Michael Selig (S) and Phillipe Giguere (G) of the University of Illinios at Urbana-Champaign, specifically for small wind turbines. They are probably the first aerofoils designed for that purpose. Their basic geometry and design parameters are given in Table 4.1 taken from Giguere and Selig [3]. Other aerofoil sections for small turbines are described by Giguere and Selig [4] and Kogaki et al. [5].

The 16% thick SG6040 is a root aerofoil, while the other three, intended for the power-extracting outer parts of the blade, have 10% thickness with varying camber. Some of the important features of the design methodology will be discussed in the next section.

4.3 Aerofoil Lift and Drag

An aerofoil of a given shape will have a lift, l, and drag, d, dependent on U_0 (which for aerofoils is equivalent to the effective velocity U_T for blade elements), c, ρ, ν, and the angle of attack α. This leads to the definition of the lift and drag coefficients as:

$$C_l = \frac{l}{\frac{1}{2}\rho U_0^2 c} \quad \text{and} \quad C_d = \frac{d}{\frac{1}{2}\rho U_0^2 c} \tag{4.1}$$

Table 4.1 Parameters of the SG family

Aerofoil	*t/c* (%)	Camber (%)	Design C_l	Design *Re*
SG6040	16	2.5	1.1	200,000
SG6041	10	2	0.6	500,000
SG6042	10	3.8	0.9	333,333
SG6043	10	5.5	1.2	250,000

and the basic formulation:

$$C_l = C_l(\alpha, Re) \quad \text{and} \quad C_d = C_d(\alpha, Re) \tag{4.2}$$

where the Reynolds number, $Re = U_0\, c/\nu$, is directly analogous to the blade element definition in Eq. 1.11. Aerofoils are also characterized by a moment or "centre of pressure" which can be important for some aeronautical applications but not for basic wind turbine analysis and design.

Note very carefully that the lift and drag that appear in (4.1) have the units of Newtons per metre (N/m). This is because an aerofoil is two-dimensional and it makes no sense to think of the lift or drag as depending on its length. One consequence of this is the appearance of the radial width of the blade element, *dr*, in Eqs. 3.10 and 3.11 for the thrust and torque in terms of C_l and C_d. *dr* is used to convert the forces in N/m to N.

The bottom plots in the three parts of Fig. 4.3 give the lift, drag and lift:drag for the SG6040 aerofoil. Consider the results for $Re = 500{,}000$ first. For a range of α, C_l is nearly linear in α. There is a branch of theoretical aerodynamics called "thin aerofoil theory" which predicts this linear dependence. However, the linearity ends before 10°, where C_d in Fig. 4.3b begins to increase rapidly; C_d is almost constant over much of the linear range in C_l. Between $\alpha = 10°$ and 15°, the flow separates from the upper surface which "stalls" the aerofoil. Notice that the maximum *l*/*d* occurs before the end of the linear region. From the previous chapter the ratio is more important for wind turbine design than the individual values of lift and drag. What is particularly noticeable about the drag results for the SG6040 (greatest thickness) and the SG6043 (greatest camber) shown at the top of Fig. 4.3 is the local maxima appearing in C_d in the otherwise low-drag region at $Re = 10^5$. This is generally the result of a laminar separation bubble, which will usually be forced to burst early by further increases in α. The most important features of laminar separation are described below and the interested reader is referred to Shyy et al. [2] for further details on the physics and calculation of laminar separation and its signature effects on aerofoil pressure distributions. Laminar separation has a major impact on the lift:drag ratio shown in Fig. 4.3c for both aerofoils. (Unfortunately pressure distributions are not available for the SG aerofoils.) The ratio depends significantly on *Re* and this dependence usually works against the designer of small wind turbines. Note also that the angle for maximum lift:drag is *Re* dependent, which means that any constant pitch turbine (as are nearly all small turbines) cannot operate at maximum efficiency as the wind speed changes if the control strategy aims to keep λ constant.

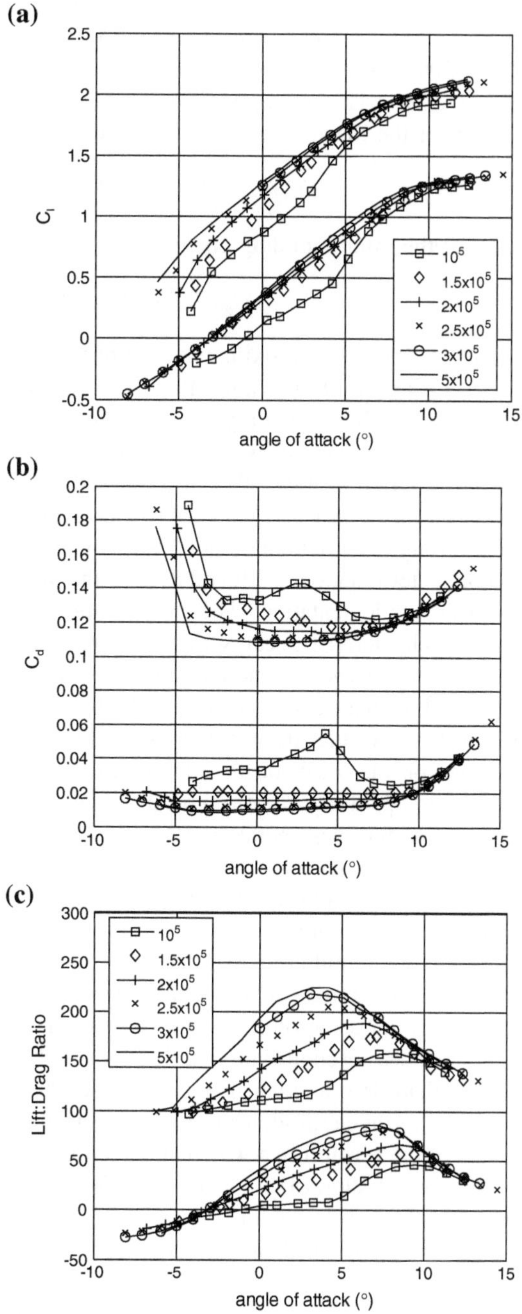

Fig. 4.3 **a** Lift coefficient of the SG6040 (*bottom*) and SG6043 aerofoils. **b** Drag coefficient of the SG6040 (*bottom*) and SG6043 aerofoils. **c** Lift:Drag ratio of the SG6040 (*bottom*) and SG6043 aerofoils. The latter is shifted upwards by 1 unit. The legend gives the Reynolds number

Figure 4.3 shows that the large change in lift:drag is due to a more gradual increase in C_l and decrease in C_d with increasing Re. This trend continues above $Re = 500{,}000$. A review of the large number of wind tunnel measurements of the NACA 0012 section, McCroskey [1] suggested the following empirical formulae for the linear part of the lift curve:

$$dC_l/d\alpha = 0.1025 + 0.00485\log_{10}\left(Re/10^6\right) \tag{4.3}$$

and for the minimum drag coefficient, C_{d0}:

$$C_{d0} = 0.0044 + 0.018\,Re^{-0.15} \tag{4.4}$$

for $Re \geq 500{,}000$ approximately. Unfortunately he did not provide a data correlation for $C_d(\alpha)$. From Eq. 4.3 the lift increases and, from (4.4), drag decreases as Re increases. It is generally the case, however, that the changes are not great in regions where relations like (4.3) and (4.4) apply. Equations 4.3 and 4.4 were used in the program LandD_0012.m described in Chap. 3. Please note that the final equation in the last part of that program where $C_d(\alpha)$ is determined from C_{d0}, is only a rough approximation developed by the author. It is used to complete a very simple determination of lift and drag so that attention can focus on the other issues of BET implementation. The next chapter shows that incorporating measured aerofoil characteristics can seriously increase the size of a blade element program.

From the answers to the Exercises in Chap. 1, it can be inferred that most large wind turbine blades operate at $Re > 500{,}000$, at least near the tip, the region where Chap. 5 shows that most power is produced. However, this Re is the *maximum* reached on the 5 kW turbine described in that chapter. From Exercise 1.10, the minimum operating Re of a typical 500 W turbine is less than 8,400. Thus many small wind turbines operate at Reynolds numbers below 10^5, the minimum of the data in Fig. 4.3. To understand the effects of low Re, it is important to be clear about the relationship between the lift and the pressure distribution around an aerofoil. Recall that there are two forces acting at any point on a two-dimensional body immersed in a fluid flow: that due to the pressure is normal to the surface and that due to the shear stress is tangential. The magnitude of the pressure is usually by far the larger. For an aerofoil, and for most streamlined bodies with no separation, the pressure does not contribute significantly to the drag. This is why the lift can be much higher than drag as seen in Fig. 4.3c. As the Re decreases, the boundary layers attached to the aerofoil surface increase in thickness and so change the effective shape of the aerofoil in a way that decreases lift and increases drag.

Figure 4.4 shows the *computed* surface pressures on a NACA0012 aerofoil at $\alpha = 0°$, 4°, and 8° obtained using the Matlab program Pablo.[1] C_p, the pressure coefficient, is defined as the gauge pressure, $P - P_0$ (where the latter is the free-stream static pressure) at the position x along the chord defined in Figs. 4.1 and 4.2, divided by the free-stream dynamic pressure:

[1] Can be downloaded from: http://www.nada.kth.se/~chris/pablo/ (accessed 4 Mar 2010).

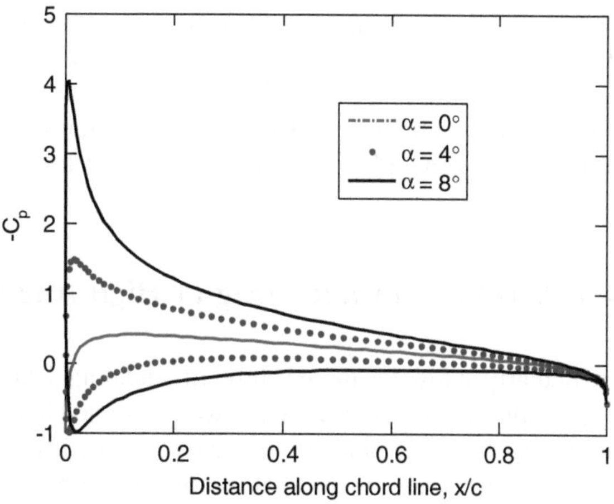

Fig. 4.4 Computed pressure distribution around a NACA 0012 aerofoil

$$C_p = \frac{P - P_0}{\frac{1}{2}\rho U_0^2} \tag{4.5}$$

and aerodynamic convention has been followed by plotting negative (lift-producing) C_p upwards. Unfortunately the pressure coefficient has the same symbol as the turbine's power coefficient, but the former is used only in this section. Note that there are two values of C_P for each *x*/*c*; the (usually) negative value for the upper surface and the other for the lower surface. The upper surface of negative gauge pressure is often called the "suction" surface and the lower one the "pressure" surface. It is easy to show that the area between the upper and lower surface values gives C_l; note the coincidence of the values for this symmetric section at $\alpha = 0°$. These simple calculations do *not* include the effects of viscosity, and so apply, in principle, only at infinite *Re*, but are sufficiently representative to indicate the acceleration of the flow over the upper surface that becomes increasingly rapid as α increases.

The boundary layers on the upper and lower surfaces begin at the stagnation point where $C_p = 1$. As *Re* decreases, these boundary layers remain laminar for a longer distance along the aerofoil surface. After the minimum pressure point on the upper surface, the boundary layer faces a region of adverse pressure gradient—where C_p is decreasing—which it is more easily withstood when turbulent at higher *Re* than when laminar at low *Re*. The usual result is separation for the low *Re* flow followed by a laminar separation bubble. Soon after the flow separates, it goes through transition to turbulence, and reattaches, often quickly. The separation region is usually identifiable on the C_p plot by its nearly constant pressure, e.g. [2].

There is very little data on aerofoil performance at $Re < 10^5$ partly because it is difficult to measure accurately the small forces involved. Laitone [6] and others suggested that the best aerofoils at $Re < 10^5$ are very thin—less than about 2.5% of the chord—and have around 5% camber. The thinness can be a major problem, because the centrifugal loads on a blade get larger as rated power decreases, see Table 1.7. In other words, it desirable for structural reasons to use thick aerofoil sections near the hub of small turbines, but this would lead to poor aerodynamic performance. Furthermore, Chap. 6 demonstrates that the hub region is especially important in generating the starting torque at low wind speed.

4.4 Aerofoil Lift and Drag at High Angles of Attack

Particularly at low *Re*, there is little lift and drag data for high angles of attack, but it is often the case that the twist of a well-designed horizontal-axis blade is close to 0° at the tip. When the blade is stationary its tip α is nearly 90° so that performance at high α is important for the analysis of starting behaviour in Chap. 6. Sheldahl and Klimas [7] measured the lift and drag of the NACA 0012 and NACA 0015 at high α as did [8] for the NACA 0012 and [9] for a number of NACA four digit aerofoils, of which the 4415 and 4418 are given here. Also shown are the lowest turbulence results for the 21% thick NACA 65_4-421 section from Devinant et al. [10]. At their maximum turbulence level of 16%, the high-α lift did not change significantly from that shown but the maximum C_d increased to 2.2. The data for $Re > 2 \times 10^5$ are collected in Fig. 4.5 and the data sources are listed in Table 4.2. Figure 4.6 has results for lower *Re* with the sources listed in Table 4.3. Low *Re* data comes from [11] for the NACA 0012 and [5] for the MEL 081 a 14% thick section specially designed for small wind turbines. At high α any reasonably thin aerofoil of small camber should behave more or less as a two-dimensional, thin flat plat held normally to the flow for which $C_l \approx 0$ and $C_d \approx 2$. Therefore, the flat plate data [12][2] are included. Also shown in the figures are the two equations

$$C_l = A \sin 2\alpha \tag{4.6}$$

and

$$C_d = B - C \cos 2\alpha \tag{4.7}$$

where *A*, *B*, and *C* can be *Re* and aerofoil dependent [13]. Figures 4.5 and 4.6 show Eqs. 4.6 and 4.7 respectively with typical values of $A = B = C = 1$. As an aside, the flat plate is an obvious counter-example to the simple-minded but surprisingly common explanation of aerofoil lift purely as a consequence of Bernoulli's equation and the supposed fact that the air must speed up as it flows over the upper part of the section. There can be no speeding up over a thin flat plate.

There is more scatter in the low-*Re* data and the apparent trend of decreasing C_l and C_d as *Re* decreases is not consistent. More data is urgently needed.

[2] The data is taken from Table X of the paper. In the appendix the authors suggest that the results for $\alpha = 90°$ should be reduced by 13.5% and those at 30° by 8% to account for wind tunnel blockage. The correction at other angles was found by linear interpolation.

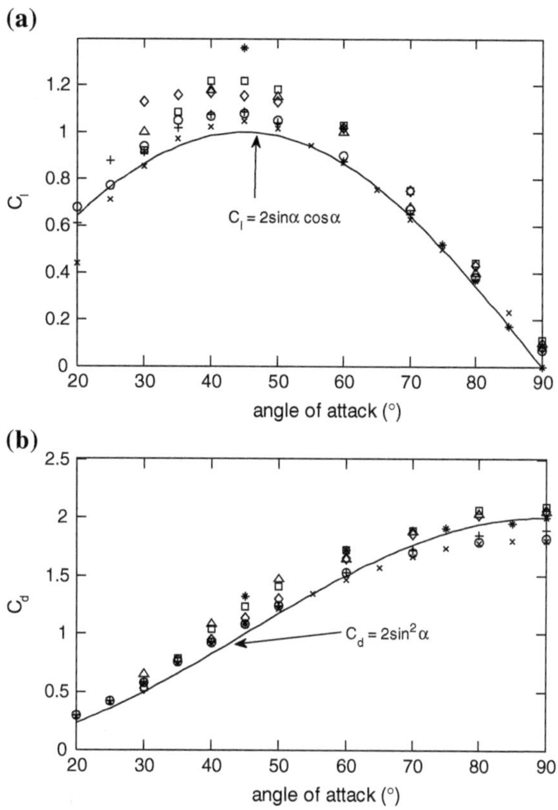

Fig. 4.5 **a** High-angle lift coefficients for Reynolds numbers above 2×10^5. **b** High-angle drag coefficients for Reynolds numbers above 2×10^5. Data sources listed in Table 4.2

Table 4.2 Source of data for Fig. 4.5

Symbol	Source	Aerofoil	*Re*
+	Michos et al. [8]	NACA 0012	7.6×10^5
O	Sheldahl and Klimas [7]	NACA 0012	5×10^5
×	Sheldahl and Klimas [7]	NACA 0015	5×10^5
Δ	Devinant et al. [10]	NACA 65_4-421	4×10^5
□	Ostawari and Naik [9]	NACA 4415	2.5×10^5
◇	Ostawari and Naik [14]	NACA 4418	2.5×10^5
*	Bruining [15]	Cambered flat plate	2×10^5

Another way to view the data is to assume that the lift and drag arise almost entirely from the pressure distribution on a thin flat plate or aerofoil at high angles. Then the lift:drag ratio must equal $1/\tan\alpha$. A test of this is shown in Fig. 4.7 which shows significant departures at low *Re*. The decrease in C_l with decreasing *Re* at moderate α seems, therefore, to be more significant than that in C_d. The increasing viscous stress on the aerofoil as *Re* decreases will tend to increase the drag and decrease the lift.

Table 4.3 Source of data for Figs. 4.6 and 4.7

Symbol	Source	Aerofoil	*Re*
+	Fage and Johansen [12]	Flat plate	1.6×10^5
×	Kogaki et al. [5]	MEL 081	10^5
*	Bruining [15]	Cambered flat plate	10^5
∇	Bruining [15]	Cambered flat plate	6×10^4
O	Zhou et al. [11]	NACA 0012	5.1×10^4
◇	Zhou et al. [11]	NACA 0012	1.05×10^4
□	Zhou et al. [11]	NACA 0012	5.3×10^3

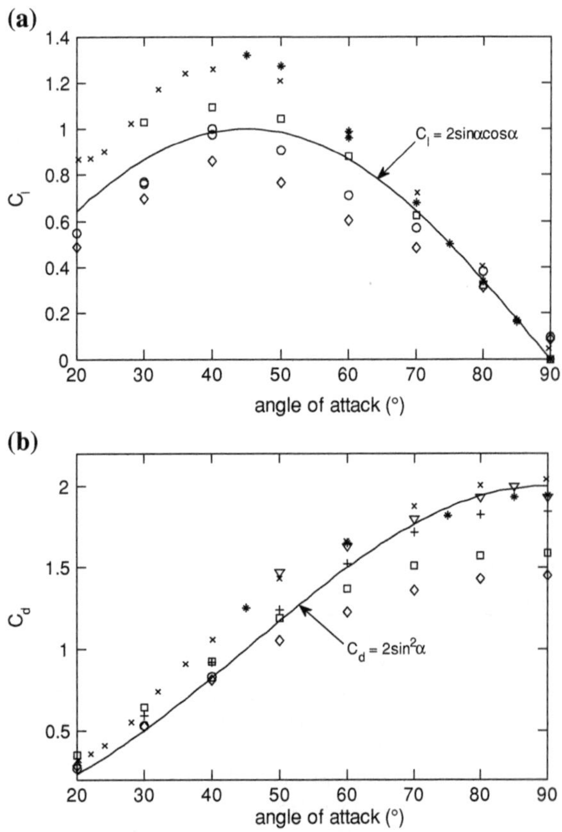

Fig. 4.6 **a** High-angle lift coefficient for Reynolds numbers below 2×10^5. **b** High-angle drag coefficient for Reynolds numbers below 2×10^5. Data sources listed in Table 4.3

4.5 The Circulation

Section 3.4 considered the two-dimensional analogue of the axisymmetric blade elements at radius r. This analogue is sketched in Fig. 4.8, where only four of the infinite cascade of aerofoils (up and down the page) is shown. In terms of the blade element chord, c, the non-dimensional spacing between the elements is just the inverse of the solidity, σ, defined by Eq. 3.14. Now assume for simplicity that the

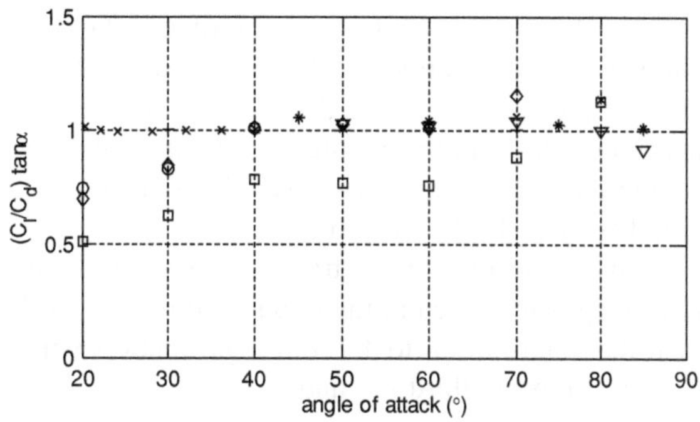

Fig. 4.7 Relationship between C_l and C_d at high angles and Reynolds number below 2×10^5. Data sources listed in Table 4.3

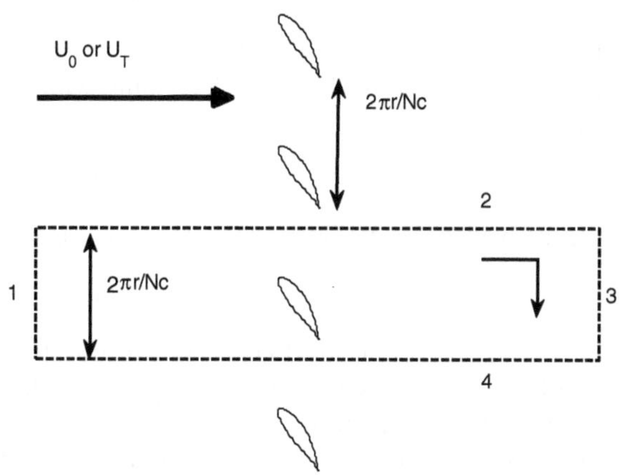

Fig. 4.8 Circulation around a cascade of blade elements at radius *r*. The four legs of the rectangular circuit used to determine the circulation are numbered

aerofoil produces only lift, which must have a component up the page to produce torque to rotate the blades. To determine the blade circulation, Γ, the clockwise circuit around the rectangle in Fig. 4.8 is divided into four legs. The analysis is made easier if the radius r is also the radius of the streamtube intersecting that element. As a result, legs 1 and 3 must be close to the blade.

The circulation is defined for any closed contour by the line integral of the velocity:

$$\Gamma = \oint \mathbf{U} \cdot d\mathbf{l} \tag{4.8}$$

where $\mathbf{U}$ is the velocity vector, and $d\mathbf{l}$ is the increment along the curve, and has the units of velocity × length or m^2/s in the SI system. Thus the circulation around a circular contour downstream of the blades and centred on the turbine's axis (in a plane parallel to the rotor disk) is $2\pi rW$ where W is the circumferential velocity. This shows the close connection between Γ and angular momentum. One of the other very useful properties of Γ is that by Gauss' theorem it is equal to the area integral of the vorticity within the contour, which is normally confined to the boundary layers on the upper (suction) and lower (pressure) surfaces of the blade.

The first leg in Fig. 4.8 is upstream of the blades. As it is also a component of a circular contour centred on the turbine axis, the first leg cannot contribute to Γ because the wind is assumed steady, one-dimensional and spatially uniform, and hence inviscid, and the circulation can never be changed in an inviscid fluid. (This simple and elegant argument was devised by the famous British aerodynamicist G.I. Taylor in 1915, see [16].)

Now the second leg is traversed from upstream to downstream whereas the fourth leg is traversed in the opposite direction. If the width of the contour is σ^{-1}, then the contribution to Γ from legs 2 and 4 will cancel. Thus Γ is determined entirely from the third leg and so

$$N\Gamma = 2\pi r W_2 \tag{4.9}$$

where W_2 is the average circumferential velocity in the wake. Note that this equation is identical to Eq. 2.8. Combining with (3.5) and (3.11) for zero drag, gives

$$L = \rho U_T \Gamma \tag{4.10}$$

If the blades are stationary, then $U_T \rightarrow U_0$ as $\sigma \rightarrow 0$, and

$$L = \rho U_0 \Gamma \tag{4.11}$$

which is the famous Kutta-Joukowski equation for aerofoils. Circulation around aerofoils manifests itself as the so-called upwash upstream and the downwash downstream of the aerofoil.

It is interesting that the derivation of (4.11) for aerofoils is much more difficult than this "proof" for a cascade; see, for example, Sect. 6.4 of [17]. There is, however, one major difference between aerofoils and blade elements of finite solidity: whereas no circumferential velocity can be "induced" upstream of blades, there is no such constraint on aerofoils or indeed, on a cascade of aerofoils. In other words, there can be no net upwash for wind turbine blades. The assumption that half the far-wake circumferential velocity defines the value of a′ for the velocity triangle in Fig. 3.2, must, therefore, be viewed as an attempt to redistribute W so as to produce an upwash for the aerofoils comprising the blade elements. Aerofoil behaviour can, therefore, only be an approximation for blade elements at sufficiently small values of W in the wake; typically W increases as σ increases.

In addition to its fundamental connection with angular momentum, two important features of Γ are that it measures vortex strength and it is a conserved quantity in an otherwise inviscid flow; if the blades have a "bound" vorticity of strength Γ, then the strength of the vortices trailing from the blades is also Γ. Note that it is the circulation that is conserved, not the vorticity. The significance of conservation of circulation is that the simplest wake structure for a wind turbine, with a uniform U_∞ and leading to the Betz–Joukowsky limit, is when the N hub vortices lie along the turbine axis and the N tip vortices are constant diameter helices in the far-wake. The trailing helical vortices in this so-called Joukowsky wake, are analogous to those behind an aircraft wing. The effects of helical trailing vorticity on the flow over the blades is, however, much more difficult to calculate than the effects of the nearly straight wing tip vortices.

One way to achieve this wake structure, and, therefore, to have a turbine whose performance can approach the Betz–Joukowsky limit, is with constant bound circulation of the blades. Combining Eq. 4.9 with Eqs. 3.5 and 3.11, and ignoring the drag in the latter, leads to

$$\Gamma \approx \frac{1}{2} U_T c_l c \tag{4.12a}$$

For sufficiently high values of λ, $U_T \approx \lambda r$, and

$$\Gamma \approx \frac{1}{2} \lambda r c_l c \tag{4.12b}$$

so that if C_l remains roughly constant, c must decrease with radius, as shown in Table 3.1. This decrease is a feature of all well-designed wind turbine blades. Furthermore, the higher the value of λ, the smaller c needs to be. It is also a feature of efficient blades that solidity decreases with increasing optimal λ. Equation 4.12 is important also because it links the aerofoil properties of the blade elements with the strength of the trailing vortices. In the next chapter this link is developed further to include U_∞, the velocity in the far-wake. Since $U_\infty = 1/3$ for the Betz–Joukowsky limit, considering the circulation leads to specific equations for the chord and twist of a maximum-power-producing blade, the first major goal of the blade designer. These equations also specify the value of the optimum circulation.

4.6 Further Discussion on Reynolds Number, High Incidence, and Aspect Ratio

The aerofoil data presented so far in this chapter have demonstrated graphically the way that lift and drag can alter with angle of attack and Reynolds numbers. Typical values of *Re* for small wind turbines were documented in Chap. 1, and the significance of *Re* for power maximisation is discussed in Chap. 7. Here is a good place, however, to continue the process of familiarisation with *Re* effects. In addition, it is important to introduce the major differences between aerofoils and three-dimensional lifting bodies in terms of the aspect ratio.

Figure 4.9 shows an idealised operating trajectory of the 500 W 1.94 m diameter three-bladed turbine described in detail in Chaps. 6 and 9 in terms of α with *Re* for the tip and the hub; note that *Re* is plotted logarithmically. The curves labeled "starting" were calculated for the blades starting from rest and a wind speed of 3 m/s. Starting is completed when the turbine reaches its operational λ of 7 and power production commences. Then α does not change as long as the controller keeps λ nearly constant

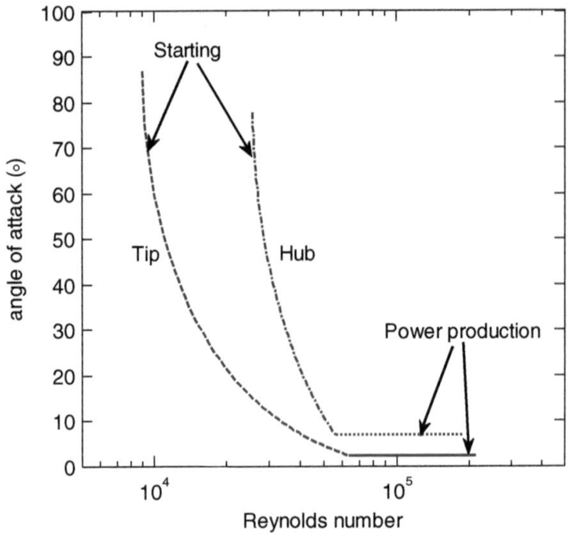

Fig. 4.9 Approximate operating trajectory of a 500 W wind turbine

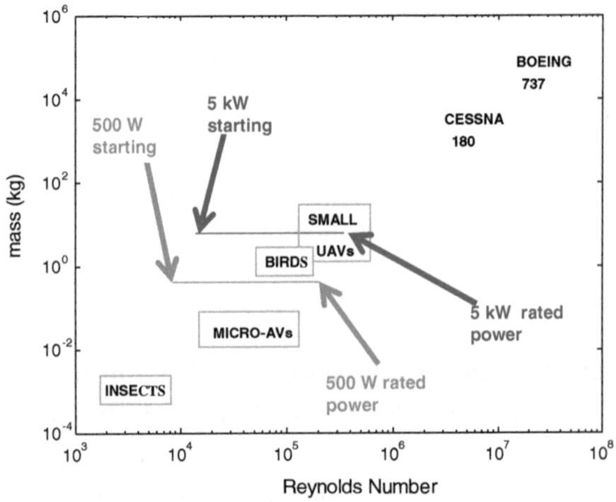

Fig. 4.10 Reynolds number ranges for small wind turbines and other aerodynamic bodies

for maximum power. For the reasons given in Chaps. 6 and 7, starting is far more important to small wind turbines than stopping which is not shown in Fig. 4.9. Also missing is any indication of the stationary blades at the extreme wind speed, one of the load cases considered in Chap. 9. Nevertheless, it is clear that high angles on small wind turbine blades are associated mainly with low *Re*.

Figure 4.10, based on Fig. 1 of [18], shows the tip Reynolds number range for the 500 W turbine and the Aerogenesis 5 kW turbine from Chap. 1 in comparison with a range of flight "vehicles". Micro-aerial vehicles (Micro-AVs) and unpiloted aerial vehicles (UAVs) are commercial and military versions of model aircraft. The use of blade mass for the vertical scale is not strictly useful for wind turbines but it does separate the data. The range of *Re* for wind turbine blades comes about solely from the change in velocity; it was assumed that the starting velocity (for a stationary blade) is 3 m/s, and the rated speed is 10 m/s. At the latter speed, the tip speed ratio, λ, is around 7, so the tip velocity is over 70 m/s, giving a 20:1 range in *Re*. The figure also shows that most commercial aviation occurs at $Re > 10^6$—the Cessna 180 is a single engine, four-seater aircraft—and that the lower limit of $Re = 10^5$ for the measured aerofoil data for the SG6040, SG6041, and SG6043, is too high for many small wind turbine applications.

Mueller and DeLaurier [16] explain that aerofoil performance below about $Re = 500{,}000$ is governed primarily by effects of the laminar separation bubble that forms on the upper surface as mentioned in Sect. 4.3. (In this general description of Reynolds number effects, it must be appreciated that the actual values are not precise and will be shape dependent.) As *Re* increases from 200,000 to about 500,000, the bubble gets shorter and the drag it causes reduces, leading to higher l/d. Between about 70,000 and 200,000 it is possible to achieve laminar flow without a bubble, which can lead to impressive performance, as shown in Fig. 4.3 for the SG6043 at $Re = 200{,}000$. Note that each of the SG6040, SG6041, and SG6043, suffer from obvious effects of laminar separation bubbles, at their lowest measured *Re*. Between 30,000 and 70,000, aerofoil thickness has a direct influence on bubble formation, which is why thin sections behave better in this region. Below about 50,000, transition in the separated flow may not occur before the trailing edge and there is no reattachment.

It is again emphasised that aerofoils are two-dimensional and that the conversion of aerofoil lift and drag (in N/m per unit width) into actual forces (in N) on an aerodynamic body involves considering the three-dimensional effects, which usually scale with the aspect ratio, *AR*, defined as the planform area of a blade or wing divided by its span. For a rectangular wing of span b and chord c, $AR = b/c$. Thus $AR \to \infty$ is the aerofoil "limit". Most basic aeronautics texts, such as Bertin and Cummings [19], give the following formula for the total drag on a rectangular (or similarly shaped) wing:

$$C_D = C_{D0} + \frac{C_L^2}{\pi(AR)e} \tag{4.13}$$

where the use of capital letters for the subscripts emphasise that the lift and drag are of three-dimensional bodies. C_{D0} is the aerofoil drag converted into a three-dimensional value by multiplying both the drag and the denominator of Eq. 4.1 by the span, and the second term on the right is often called the induced drag. e is the "span efficiency factor" and has a value of 0.8–0.9 for most wings. For typical C_L values of around unity, the induced drag dominates at sufficiently low *AR*. However, *AR* for most wind turbine blades is in the range 10–30, and the calculations of blade circulation to be discussed in Chap. 5, suggest a much more uniform distribution than the elliptical loading on wings that underlies Eq. 4.13. This equation probably over-estimates *AR* effects for wind turbine blades, and is not used.

One area where *AR* effects can be important is at high incidence. As demonstrated in Sect. 4.4, aerofoils should behave as a thin flat plate when $\alpha \to 90°$

and it is well known that *AR* partially determines the drag. This obviously cannot be a lift-induced effect as there is very little, if any, lift when $\alpha = 90°$. A data correlation that reproduces the flat plate results in Table 7.2 of [20] was apparently proposed by Hoerner:

$$\begin{aligned} C_{D,\max} &= 1.11 + 0.018(AR) \quad \text{for } AR < 50 \\ &= 2.0 \quad\quad\quad\quad\quad\quad\;\; \text{for } AR \geq 50 \end{aligned} \tag{4.14}$$

Based on this Viterna and Corrigan proposed—see [21],

$$C_D = C_{D,\max} \sin^2 \alpha + B_2 \cos \alpha \tag{4.15a}$$

where

$$B_2 = \left(C_{D,s} - C_{D,\max} \sin^2 \alpha_s\right) / \cos \alpha_s \tag{4.15b}$$

and

$$C_L = \frac{1}{2} C_{D,\max} \sin 2\alpha + A_2 \cos^2 \alpha / \sin \alpha \tag{4.16a}$$

where

$$A_2 = \left(C_{L,s} - C_{D,\max} \sin \alpha_s \cos \alpha_s\right) \sin \alpha_s / \cos^2 \alpha_s \tag{4.16b}$$

and the subscript "s" indicates a matching point at low α, often within the stall region. Despite the wide use of these equations, the author believes that they are more complex than is warranted by the data, as demonstrated in Figs. 4.5 and 4.6, which suggest the effects of *Re*, aerofoil geometry, and possibly the variations in turbulence levels, wind tunnels, and measuring techniques. In addition, Chap. 3 noted that blade element theory assumes the independence of the elements. If this is correct then the forces on each element cannot depend on the aspect ratio of the blade. Chapter 6 analyses turbine starting, which involves high angles of incidence, using Eqs. 4.6, 4.7, and 4.14–4.16 and compares the predictions to measurements.

While moderate aspect ratios are not of much interest for small wind turbines, small values are. Chapter 7 shows that the prototype shapes of most tail fins are slender delta wings or low *AR* rectangles. That chapter considers their lift and drag in considerable detail.

4.6.1 Further Reading

A good description of the early NACA work on aerofoil design is found in Abbott and von Doenhoff [22]. All the NACA and NASA reports referred to in this book can be downloaded from the NASA Technical Report Server website http://naca.larc.nasa.gov/search.jsp. General information on the fundamentals of aerofoils with aeronautical application can be found in books such as Bertin and

Cummings [19]. Lift and drag of aerofoils can be found in Althaus [23], Althaus and Wortmann [24], and Miley [25] and from the web site maintained by Professor Michael Selig's group at the University of Illinois, Urbana-Champaign http://www.ae.illinois.edu/m-selig/ (accessed 1 Oct 2010). Hard copies of their data can be found in Selig et al. [26, 27] and Lyon et al. [28]. The small wind turbine community owes a great debt to NASA and Professor Selig for making available so much valuable information.

4.6.2 Exercises

1. If Figs. 4.1 and 4.2 are thought of as sections through typical wind turbine blades, will the wind direction be up or down the page? What is the direction of rotation?
2. The three blades of a vertical axis wind turbine are attached to the hub by horizontal support arms, one for each blade. The turbine is 4 m in diameter and the chord length of the non-lifting supports is 7 cm. To determine the effect of the supports' drag on performance, the blades were removed and the main shaft was instrumented to measure the torque, Q, necessary to drive the supports at an angular velocity Ω (in rad/s) in still air. The data were fitted by the curve Q (Nm) $= 0.504\ \Omega^2$. Show that this is consistent with the drag coefficient C_d being independent of radius and determine its value.
3. Show that the pressure coefficient, C_P, has the value of unity at the stagnation point of an aerofoil in incompressible flow.
4. Why can C_P never exceed this value?
5. Show that Eqs. 4.6 and 4.7 with $A = B = C = 1$ imply that the fluid force on a thin flat plate is due entirely to the pressure at high α.
6. Show that (4.6) and (4.7) are consistent with $C_l/C_d = 1/\tan\alpha$ in general only when $A = B = C$.
7. Using your fluid dynamics textbook, review the discussion on the origin of lift in fluid flow in terms of Bernoulli's equation and circulation and the relationship to the material in this chapter.
8. Go the web site of a large wind turbine manufacturer to see if you can find enough data to include the operational range of Reynolds number on Fig. 4.10. Note that blade mass has no direct operational consequences and is used mainly to separate the data.
9. The Lock number is used in helicopter aerodynamics to measure the ratio of aerodynamic to inertial blade loads. At high lift, the Lock number can be approximated as

$$\rho C_{l,\alpha} c R^4 / J$$

where ρ is the air density, $C_{l,\alpha}$ is the slope of the lift coefficient (assumed constant for the linear range), and J is the inertia. For a 5 kW, 2.5 m long blade,

with $J = 8$ kg m^2, estimate the Lock number and compare the value with the range of 5–10 which is common for helicopters.
10. Review the statements in Sect. 1.8 regarding the R-dependence of turbine inertia in the light of the Lock number.
11. Explain why optimum blade element performance occurs when the lift:drag ratio is maximised rather than, say, where lift is maximised?
12. Explain why the centre of pressure is not important for blade element analysis.

References

1. McCroskey W (1987) A critical assessment of wind tunnel results for the NACA 0012 airfoil. NASA Technical Memorandum 100019
2. Shyy W, Lian Y, Tang J, Viieru D, Liu H (2008) Aerodynamics of low Reynolds number flyers. Cambridge University Press, New York
3. Giguere P, Selig MS (1998) New airfoils for small horizontal axis wind turbines. J Solar Energy Eng 120:108–114
4. Giguere P, Selig MS (1997) Low Reynolds number airfoils for small horizontal axis wind turbines. Wind Eng 21:367–380
5. Kogaki T, Matsumiya H, Iida M, Inaba T, Yoshimizu N, Kieda K (2002) Development and experimental verification of an airfoil for small wind turbines. In: Proceedings of 2002 global windpower conference and exhibition
6. Laitone E (1997) Aerodynamic lift at Reynolds numbers below 70,000. AIAA J 34:1941–1942
7. Sheldahl R, Klimas P (1981) Aerodynamic characteristics of seven symmetrical airfoil sections through 180-degree angle of attack for use in aerodynamic analysis of vertical axis wind turbines, Sandia National Laboratories, report SAND80-2114. http://windpower.sandia.gov/abstracts/802114A.pdf. (accessed 5 Aug 2010)
8. Michos A, Bergeles G, Athanassiadis N (1983) Aerodynamic characteristics of NACA 0012 airfoil in relation to wind generators. Wind Eng 7:247–262
9. Ostowari C, Naik D (1984) Post stall studies of untwisted varying aspect ratio blades with a NACA4415 airfoil section—part I. Wind Eng 8:176–194
10. Devinant P, Laverne T, Hureau J (2002) Experimental study of wind-turbine airfoil aerodynamics in high turbulence. J Wind Eng Ind Aerodyn 90:689–707
11. Zhou Y, Md Mahbub A, Yang HX, Guo H, Wood DH (2010) Fluid forces on a very low Reynolds number airfoil and their prediction. Int J Heat Fluid Flow 32:329–339
12. Fage AR, Johansen FC (1927) On the flow of air behind an inclined flat plate of infinite span. Proc R Soc A 116:170–197
13. Wang ZJ (2005) Dissecting insect light. Annu Rev Fluid Mech 37:183–210
14. Ostowari C, Naik D (1985) Post Stall Studies of Untwisted Varying Aspect Ratio Blades with a NACA44XX Airfoil Sections –Part II. Wind Engineering 9:149–164.
15. Bruining A (1979) Aerodynamic Characteristics of a Curved Plate Airfoil Section at Reynolds Numbers 60,000 and 100,000 and angles of attack from -10 to +90 degrees, Delft University of Technology, Report LR-281 http://repository.tudelft.nl/view/ir/uuid:6b92442a-01f7-4b7c-8d53-c4f10720ff3e/ (accessed 26 Sept, 2010)
16. Taylor GI (1963) The "rotational inflow factor" in propeller theory. In: G. Batchelor (ed) The scientific papers of G.I. Taylor CUP 3:59–65
17. Batchelor G (1967) An introduction to fluid dynamics. Cambridge University Press, Cambridge

18. Mueller T, DeLaurier J (2003) Aerodynamics of small vehicles. Annu Rev Fluid Mech 35:89–111
19. Bertin J, Cummings R (2009) Aerodynamics for engineers, 5th edn. Pearson Prentice Hall, Upper Saddle River
20. White F (2011) Fluid mechanics, 7th edn. McGraw Hill, New York
21. Amarante Mesquita ALA, Alves ASG (2001) An improved approach for performance prediction of HAWT using strip theory. Wind Eng 24:417–430
22. Abbott I, von Doenhoff A (1959) Theory of wing sections. Dover, New York
23. Althaus D (1996) Niedriggeschwindigkeitsprofile: Profilentwicklungen und Polarenmessugen im Laminarwindkanal des Instututs für Aerodynamik und Gasdynamik der Universität Stuttgart. F. Vieweg, Braunschweig
24. Althaus D, Wortmann FX (1981) Stuttgarter Profilkatalog. F. Vieweg, Braunschweig
25. Miley SJ (1982) A catalog of low Reynolds number airfoil data for wind turbine applications. Report prepared under U.S. DoE Wind Energy Program
26. Selig MS, Donovan JF, Fraser D (1989) Airfoils at low speeds. Soartech, Virginia Beach
27. Selig MS, Lyon CA, Giguere P, Ninham CP, Guglielmo JJ (1996) Summary of low-speed airfoil data, vol 2. Soartech, Virginia Beach
28. Lyon CA, Broeren AP, Giguere P, Gopalarathnam A, Selig MS (1997) Summary of low-speed airfoil data, vol 3. Soartech, Virginia Beach

Chapter 5
Blade Element Calculations

5.1 Introduction

Equations 3.10 and 3.11 are the basic blade element equations and there probably have been many programs written to implement them. Two important and publically-available examples are PROPID[1] written by Professor Michael Selig at the University of Illinois, Urbana-Champaign and WT_Perf, written by Marshall Buhl at NREL.[2]

Over the years, a large number of modifications to the basic equations have been proposed. The only one considered here is meant to account for the finite number of blades ($N < \infty$) on any real turbine. The necessity of some correction for finite N comes from realising that the streamtube analysis of Chaps. 2 and 3 assumed that the velocities and pressures are uniform in the circumferential direction, whereas non-uniformities must arise for a finite number of blades. In other words, the axial velocity at the blade element may be different from U_1 which is the streamtube's average velocity. A simple and commonly used correction for this effect is "Prandtl's tip loss factor", F, defined as the ratio of the average induction factor, a, to the value at the blades, a_b: $F = a/a_b$. The implementation of the tip loss factor as described here follows closely the work of Shen et al. [1], Xudong et al. [2] and Clifton-Smith [3]. In its simplest form, the equation for F is:

$$F = 2\cos^{-1}\left(e^{-f}\right)/\pi \tag{5.1}$$

where

$$f = N(R - r)/(2\,r\sin\phi) \tag{5.2}$$

In practice, F, which is always less than unity, makes a difference of around 5–10% to the predicted turbine performance and is often neglected, especially

[1] http://www.ae.illinois.edu/m-selig/propid.html (accessed 25 Sept 2010).

[2] http://wind.nrel.gov/designcodes/simulators/wtperf/ (accessed 25 Sept 2010).

D. Wood, *Small Wind Turbines*, Green Energy and Technology,
DOI: 10.1007/978-1-84996-175-2_5,

when considering optimum performance. Clifton-Smith [3] tested a number of different tip loss methods and recommended the correction of de Vries [4] which rewrites Eqs. 3.10 and 3.11 as

$$\frac{a_b F(1 - a_b F)}{(1 - a_b)^2} = \frac{\sigma C_a}{4 \sin^2 \phi} \tag{5.3}$$

and

$$\frac{a'_b F(1 - a_b F)}{(1 + a'_b)(1 - a_b)} = \frac{\sigma C_{a'}}{4 \sin \phi \cos \phi} \tag{5.4}$$

where the top line on the left hand sides represents the momentum, from Eq. 5.3, and angular momentum, Eq. 5.4, in the wake and so involves a' and a whereas the denominators relate to the forces at the blade elements and so involve a_b and a'_b. Clifton-Smith [3] used the blade element version of the high thrust correction to the momentum equation, Eq. 2.20, so that (3.13) is replaced by

$$f_a = \left\{ \begin{array}{ll} a_b F(1 - a_b F) & a_b \leq a_c \\ a_c^2 F^2 + (1 - 2a_c F) a_b F & a_b > a_c \end{array} \right\} \tag{5.5}$$

where the usual value for a_c is 1/3. In other words, the high thrust correction begins at the Betz–Joukowsky limit. Shen et al. [1] defined two intermediate functions, Y_1 and Y_2 (in the form to be used with Eqs. 5.3 and 5.4):

$$Y_1 = 4F \sin^2 \phi / (\sigma C_a) \tag{5.6a}$$

and

$$Y_2 = 4F \sin \phi \cos \phi / (\sigma C_{a'}) \tag{5.6b}$$

which are used in the following expressions for the induction factors:

$$a_b = \frac{2 + Y_1 - \sqrt{4Y_1(1 - F) + Y_1^2}}{2(1 + FY_1)} \tag{5.7a}$$

and

$$a'_b = \frac{1}{(1 - aF)Y_2/(1 - a) - 1} \tag{5.7b}$$

5.2 Altering the Programs from Chap. 3

The chord and twist distribution shown in the file tcdist.m from Chap. 3 are fitted curves to the tabulated data in Anderson et al. [5]. They measured the power and thrust of a 3 m diameter turbine whose two blades had a NACA 4412 profile. There are two important aspects of this turbine to be noted and remembered: firstly

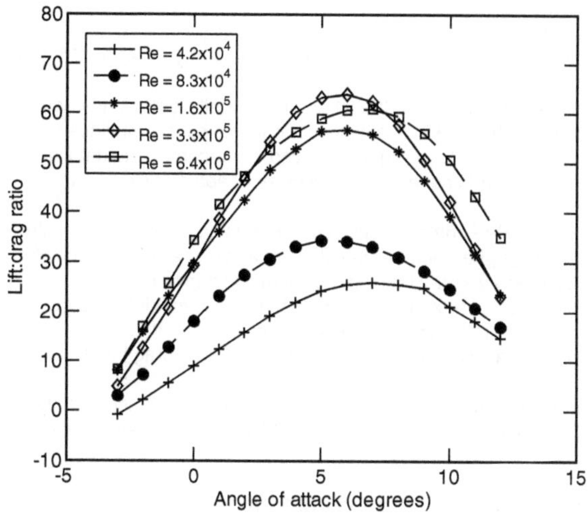

Fig. 5.1 Lift:drag ratio for the NACA 4412 aerofoil

it achieved a high maximum C_P with an old aerofoil section, so it forms a good basis for investigating the practical optimisation of blade design. Secondly, the measurements are still among the most detailed available; presumably for that reason they are considered extensively in Chap. 5 of Spera [6]. The main drawback in the current context is that the only lift and drag data for the NACA 4412 section were obtained in 1937 in a wind tunnel with high free-stream turbulence (about 2%), so their accuracy is questionable. On the other hand, the lift and drag were measured for $42{,}000 \leq Re \leq 640{,}000$, which is a very appropriate range for this wind turbine. The lift and drag data were taken from Miley [7]. The lift:drag ratio is shown in Fig. 5.1 as function of α and Re. The data, as are all data and programs mentioned in this chapter, are available from the online materials (http://extras.springer.com).

The tabulation of lift and drag data in terms of α is much more common than the data fits that were used for the NACA 0012 section in the file LandD_0012.m described in Chap. 3. To make interpolation easy, the lift and drag should be measured at the same angles for each Re value, but this does not commonly occur. For example, the drag for the NACA 4412 was not measured at 11° and 12° at the lowest Re. This is the typical situation in using aerofoil data for wind turbine calculations. The missing values were assumed to be equal to those at the nearest Re.

It is noted in passing that the lift and drag show the general trend of Reynolds number dependence that is to be expected from knowledge of Chap. 4; C_l tends to decrease and C_d increases as Re decreases, so the lift:drag ratio is strongly Re-dependent as shown in Fig. 5.1.

Linear interpolation is used to determine lift and drag as a function of α and linear interpolation in log(Re). To do this LandD_0012.m from Chap. 3 is replaced by ReadIn_4412.m and LandD.m. The first of these functions must be rewritten for every new aerofoil used but the second is general. ReadIn_4412.m reads the data from the data file naca4412.in.

```
functionReadIn_4412
global num_Re num_aoa Re_array aoa_array Cl_array Cd_array

% Function to read in the lift and drag coefficients of any aerofoil
% when the data is available in a data file containing lines:
% aoa  Cl  Cd for a range of Reynolds numbers. The data is
% assumed to be in ascending order of aoa and Re.

% num_Re - number of values of Reynolds number in data file.
% num_alpha - number of values of angle of attack at each Re.
% Re_array -array holding the Reynolds numbers.
% Re_log -    array holding the log of the Reynolds numbers
% alpha_array - array holding the angles of attack.
% Cl_array, Cd_array - arrays holding lift and drag coefficients.

% Values specific to the NACA 4412 data. This function makes use
% of the equal number of Cl and Cd values at the same aoa for each
% Re

num_Re = 5; num_aoa = [16 16 16 16 16];
max_num_aoa=max(num_aoa);
Re_array = [42000 83000 160000 330000 640000];
in = load('naca4412.in'); % This is the data file.

% End of specific data for this aerofoil. The remainder of this function
% is the same for any aerofoil

aoa_in=in(:,1); % Angle of attack is in the first column
Cl_in=in(:,2);  % Lift coefficient is in second column
Cd_in=in(:,3); % Drag coefficient is in second column
k=1;
aoa_array=zeros(num_Re, max_num_aoa);
Cl_array=zeros(num_Re, max_num_aoa);
Cd_array=zeros(num_Re, max_num_aoa);
for i = 1: num_Re
     for j = 1: num_aoa(i)
          aoa_array(i, j) = aoa_in(k);
          Cl_array(i, j) = Cl_in(k);
          Cd_array(i, j) = Cd_in(k);
          k = k+1;
     end
end
return
end
```

Note the use of global variables declared in the line

```
global num_Re num_aoa Re_array aoa_array Cl_array Cd_array
```

which provides a way of communicating information between the two functions without lengthening the argument list. None of these variables is used in the main program which, therefore, does not need the global statement.

The second function then interpolates that data for the input Reynolds number and angle of attack.

```
function [Cl, Cd] = LandD(aoa, Re)
global num_Re num_aoa Re_array aoa_array Cl_array Cd_array

% Cl and Cd are the lift and drag coefficients respectively
% for an angle of attack, aoa, and chord Reynolds number, Re

if aoa > min(max(aoa_array')) % Determine if aoa is within range
    fprintf(' aoa = %5.2f degrees too large \n', aoa)
   aoa = min(max(aoa_array'));
elseif aoa < max(min(aoa_array'))
   fprintf(' aoa = %5.2f degrees too small \n', aoa)
   aoa = max(min(aoa_array'));
end
if Re < Re_array(1) % Use the lowest Re data
   fprintf(' Re too small %8.3e  \n', Re)
   aoa_int = aoa_array(1,1:num_aoa(1));
   Cl_int=Cl_array(1, 1:num_aoa(1));
   Cd_int=Cd_array(1, 1:num_aoa(1));
   Cl=interp1(aoa_int(1,:),Cl_int(1,:), aoa);
   Cd=interp1(aoa_int(1,:),Cd_int(1,:), aoa);
   return
else if Re > Re_array(num_Re) % Use the highest Re data
   fprintf(' Re too large %8.3e  \n', Re)
   aoa_int = aoa_array(num_Re,1:num_aoa(num_Re));
   Cl_int=Cl_array(num_Re, 1:num_aoa(num_Re));
   Cd_int=Cd_array(num_Re, 1:num_aoa(num_Re));
   Cl=interp1(aoa_int(1,:),Cl_int(1,:), aoa);
   Cd=interp1(aoa_int(1,:),Cd_int(1,:), aoa);
   return
end

Re_log = log10(Re_array);

% The linear interpolation is in log(Re) and alpha

Re_in_log = log10(Re);

j_Re=find(Re_log>Re_in_log,1); % Find the first Re > input Re
factor_Re = (Re_in_log - Re_log(j_Re-1))/(Re_log(j_Re) - Re_log(j_Re-1));

% Interpolate the Cl and Cd data in angle at the higher Re

aoa_int = aoa_array(j_Re,1:num_aoa(j_Re));
Cl_int=Cl_array(j_Re, 1:num_aoa(j_Re));
Cd_int=Cd_array(j_Re, 1:num_aoa(j_Re));
Cl_up=interp1(aoa_int,Cl_int, aoa); % Use linear interpolation
Cd_up=interp1(aoa_int,Cd_int, aoa);

% Interpolate the Cl and Cd data in angle at the lower Re

aoa_int = aoa_array(j_Re-1, 1:num_aoa(j_Re-1));
Cl_int=Cl_array(j_Re-1, 1:num_aoa(j_Re-1));
```

```
Cd_int=Cd_array(j_Re-1, 1:num_aoa(j_Re-1));
Cl_low=interp1(aoa_int,Cl_int, aoa);
Cd_low=interp1(aoa_int,Cd_int, aoa);

% Now interpolate in log10(Re).

Cl = Cl_low*(1 - factor_Re) + Cl_up*factor_Re;
Cd = Cd_low*(1 - factor_Re) + Cd_up*factor_Re;
return
```

Values of *Re* passed to LandD.m that are outside the tabulated range are set equal to the closest value and a warning message is printed to the screen. The input α is first checked against the smallest of the maximum values of α measured at each *Re*:

```
if aoa > min (max(aoa_array'))% Determine if aoa is within range
```

A warning message printed to the screen if the inequality is true. A similar check is then made against the largest of the minimum values. Note that the dash (′) at the end of aoa_array′ gives the transpose which is necessary to work with the Matlab functions min and max. An input α and/or *Re* outside the tabulated range does not interrupt the program execution and it is up to the user to judge the severity of the exception(s) on the calculations.

The interpolation in angle and *Re* is done using the Matlab function interp1 using linear interpolation. The main function or program is power_calc.m which is an obvious descendent of simple_power_calc.m from Chap. 3.

```
function power_calc(Numb)
% Program to implement blade element/one-dimensional wake
% analysis for a horizontal-axis wind turbine with any number of
% blades of any length.
% The function argument and main variables are the same as in
% simple_power_calc.m described in Chapter 3.
% Note that a (the axial induction factor) used here is the value at the
% blade, a_b.

% Down to the line for delr, the program is identical to
% simple_power_calc.m from Chapter 3

delr=rad(2)-rad(1); % Determine width of blade elements

Readln_4412  % Read in Cl and Cd for the NACA 4412 section
% Readln_7062  % Read in Cl and Cd for the SD7062 section
a_c = 1/3.0; % Set a for modified thrust equations

outfile_a = fopen('out_a.dat', 'w');
outfile_b = fopen('out_b.dat', 'w');
outfile_c = fopen('out_c.dat', 'w');
heading_screen=...
' Radius iter. aoa    a    Cl    Cd   deltor    Re \n';
heading_a=...
  ' Radius iter. aoa    a    Cl    Cd     deltor    Re \n';
heading_b=' Radius   U1    gamma    deltor   delthr \n';
```

```
format_screen=' %7.4f %3d %7.2f %7.3f %7.3f %8.5f... %8.5f %8.3e \n';
format_a=' %7.4f %3d %7.2f %7.3f %7.3f %8.5f %8.5f %8.3e \n';
format_b=' %7.4f %7.5f %7.5f %8.5f %8.5f \n';
U0 = input(' Enter the wind speed in m/sec: end with -ve: ');
while U0 > 0.0;
  fprintf(outfile_a, ' For U0 = %4f m/s \n', U0);
  fprintf(outfile_b, ' For U0 = %4f m/s \n', U0);
  fprintf(outfile_c, ' For U0 = %4f m/s \n', U0);
  fprintf(outfile_c, '   TSR      Cp       Ct \n');
  lambda = input(' Enter TSR: end with -ve: ');
  while lambda > 0.0
     fprintf(outfile_a, ' For TSR = %4f  \n', lambda);
     fprintf(outfile_b, ' For TSR = %4f  \n', lambda);
     thrust = 0.0; torque = 0.0;
     fprintf('\n')
     fprintf(heading_screen)
     fprintf(outfile_a,heading_a);
     fprintf(outfile_b,heading_b);
     a = 0.3; % Initialise a
     for i = 1: nbes
        adash = 0.0; deltor = 0.0;
        lamr = lambda*rad(i);  % Local speed ratio
        sigma = 0.5*Numb*chord(i)/pi/rad(i);  % Local solidity
        for j = 1:40 % Limit the  number of iterations per b.e. to 40
           phi = atan((1 - a)/(1 + adash)/lamr);  % Eq. 3.12
           cosphi = cos(phi);
           sinphi = sin(phi);
           smallf = 0.5*Numb*(1/rad(i)-1)/sin(phi); % Eq. 5.2
           bigF = 2*acos(exp(-smallf))/pi;  % Eq. 5.1
           aoa = phi*180.0/pi - twist(i); % Find aoa
           Ut = sqrt((1-a)^2 + (lamr*(1 + adash))^2);  % Eq. 3.7a
           Re = Ut*U0*chord(i)*r_tip/visc ; % The Reynolds number
           [Cl, Cd] = LandD(aoa, Re); % Find lift and drag
           C_a = Cl*cosphi + Cd*sinphi; %  For axialforce
           C_adash = Cl*sinphi - Cd*cosphi;
           Y1 = 4*bigF*sinphi^2/sigma/C_a;  % Eq. 5.6a
           Y2 = 4*bigF*sinphi*cosphi/sigma/C_adash;  % Eq. 5.6b
           if (Y1 >= 2.0 ) % In the low thrust region
              newa = (2 + Y1 - sqrt(Y1*(4*(1-bigF) + Y1))/2 /(1+bigF*Y1);
           else % Use Glauert *`s correction for the high thrust region
              c_t = 1 -2*a_c*bigF;
              newa = (2 + Y1*c_t-sqrt((Y1*c_t+2)^2 -4*(1-a_c^2*bigF*Y1)))/2;
           end
           diffa = abs(a - newa);
           a = newa;
           adash = 1/((1-a*bigF)*Y2/(1-a)-1);  % Eq. 5.7b
           if diffa < tol*a & j > 3; break; end
        end
        aF=a*bigF; % This is a in the text
        Ngam = 0.5*Numb*Ut*Ut*chord(i)*Cl_adash/(1 - aF);
        delthr = Numb*Ut*Ut*chord(i)*delr/pi;  % Eq. 3.10
        deltor = delthr*rad(i)*C_adash;   % Eq. 3.11
        delthr = delthr*C_a; % Completing Eq. 3.10
        thrust = thrust + delthr;
        torque = torque + deltor;    % Sum the rotor torque
        fprintf(format_screen, rad(i), j, aoa, aF, Cl, Cd, deltor, Re)
```

```
        fprintf(outfile_a,format_a , rad(i), j, aoa, aF, Cl, Cd,...
                              deltor, Re);
        fprintf(outfile_b,format_b , rad(i), 1 - aF, Ngam, deltor, delthr);
    end
    cp = torque*lambda;
    fprintf(' \n')
    fprintf(' Cp = %5f,   Ct = %5f  \n', cp, thrust)
    fprintf(outfile_c,'%5f  %5f  %5f \n', lambda, cp, thrust);
        power = 0.5*cp*rho*U0^3*p *r_tip^2;
        thrust = 0.5*thrust*rho*U0^2*pi*r_tip^2;
        fprintf(' Power = %5e Watts, Thrust = %5e Newtons \n', ...
                              power, thrust)
        fprintf(' \n')
        lambda = input(' Enter TSR: end with -ve: ');
    end
    U0 = input(' Enter the wind speed in m/sec: end with -ve: ');
end
fclose(outfile_a); fclose(outfile_b); fclose(outfile_c);
```

Beside the inclusion of the tip loss and the new method of calculating a and a' based on Eqs. 5.7, the changes from simple_power_calc.m are mainly to input tabulated aerofoil data. The line

```
ReadIn_4412 % Read in Cl and Cd for the NACA 4412 section
%ReadIn_7062 % Read in Cl and Cd for the SD7062 section
```

is the only line specific to individual aerofoil sections.

The number of iterations per blade element is limited to 40. It has been found that occasionally the calculations for the first element will not converge. Usually this element contributes an insignificant amount to the power (as will be shown below) so the problem is not a major one in practice. However, it is important that the user take note of the output number of iterations when running the program to tell whether the calculations have failed to converge for any blade element.

The start of the data file naca4412.in is shown below. The first column is α in degrees, the second is C_l and the third C_d. The data down to $\alpha = 12.0°$ is for the lowest $Re = 4.2 \times 10^4$, see Fig. 5.1, and the first line of the data ($\alpha = -3.0°$) is shown for the next Re of 8.3×10^4. The rest of the file has data similarly ordered. Obviously a new aerofoil would require a new data file.

```
-3.0 -0.03 0.0353
-2.0 0.08 0.0352
-1.0 0.19 0.0345
0.0 0.30 0.0334
1.0 0.40 0.0325
2.0 0.50 0.0316
3.0 0.59 0.0308
4.0 0.67 0.0305
5.0 0.74 0.0307
6.0 0.80 0.0315
7.0 0.86 0.0331
8.0 0.91 0.0356
9.0 0.96 0.0387
10.0 0.99 0.0470
11.0 1.02 0.0568
12.0 1.03 0.0704
-3.0 0.10 0.0327
```

5.3 Running the Programs

Shown below is a typical run of tcdist.m and power_calc.m for comparison with the measurements of Anderson et al. [5] for a 3 m diameter two-bladed turbine at $U_0 = 10$ m/sec. Note that the programs are run in Matlab's command window and that >> is the prompt. The hub radius was taken as 0.133 m from Table 3.1. Fifteen blade elements were used. All blade element calculations have converged as the number of iterations is always less than 41, and no angle of attack or Reynolds number was out of range.

Figure 5.2a shows the predicted and measured C_P, and Fig. 5.2b shows the corresponding thrust coefficient. It must be noted that α was outside (higher than) the tabulated range for some blade elements for $\lambda \leq 7$ so those results must be treated with caution. Some specific results for $\lambda = 8$, 10, and 12 are shown in Figs. 5.3, 5.4, 5.5 and 5.6 inclusive. Note that the torque and thrust contributions from blade elements, given in the highlighted output below and plotted in Figs. 5.5 and 5.6, sum to $C_Q = C_P / \lambda$ and C_T respectively.

```
>> type rad_ch_tw.dat

1.5000000e+001 1.5000000e+000 1.5000000e-001
1.3000000e-001 1.6874554e-001 2.4727410e+001
1.9000000e-001 1.3699396e-001 1.6765225e+001
2.5000000e-001 1.1239713e-001 1.1276093e+001
3.1000000e-001 9.3604305e-002 7.6139593e+000
```

```
3.7000000e-001 7.9412182e-002 5.2343789e+000
4.3000000e-001 6.8764925e-002 3.6945145e+000
4.9000000e-001 6.0754160e-002 2.6531376e+000
5.5000000e-001 5.4618973e-002 1.8706279e+000
6.1000000e-001 4.9745912e-002 1.2089738e+000
6.7000000e-001 4.5668986e-002 6.3177214e-001
7.3000000e-001 4.2069663e-002 1.6519973e-001
7.9000000e-001 3.8776877e-002 -2.5840021e-001
8.5000000e-001 3.5767019e-002 -6.8200015e-001
9.1000000e-001 3.3163942e-002 -1.1056001e+000
9.7000000e-001 3.1238963e-002 -1.5292000e+000
```

```
>> power_calc(2)
Enter the wind speed in m/sec: end with -ve: 10
Enter TSR: end with -ve: 8
Radius iter. aoa   a     Cl    Cd      deltor  Re
0.1300 6   7.92  0.239  1.040 0.01878 0.00091 2.380e+005
0.1900 5   8.40  0.236  1.068 0.02034 0.00140 2.460e+005
0.2500 7   8.82  0.238  1.092 0.02185 0.00187 2.492e+005
0.3100 8   9.07  0.237  1.106 0.02290 0.00231 2.488e+005
0.3700 8   9.07  0.232  1.105 0.02291 0.00272 2.470e+005
0.4300 7   8.85  0.224  1.093 0.02200 0.00311 2.455e+005
0.4900 6   8.52  0.217  1.075 0.02081 0.00348 2.452e+005
0.5500 7   8.19  0.212  1.056 0.01958 0.00384 2.460e+005
0.6100 7   7.92  0.209  1.040 0.01867 0.00420 2.474e+005
0.6700 7   7.70  0.206  1.026 0.01815 0.00453 2.486e+005
0.7300 7   7.50  0.203  1.013 0.01764 0.00483 2.489e+005
0.7900 8   7.32  0.198  1.001 0.01723 0.00507 2.478e+005
0.8500 8   7.18  0.193  0.992 0.01691 0.00522 2.456e+005
0.9100 9   6.98  0.187  0.978 0.01646 0.00520 2.434e+005
0.9700 11   6.35  0.176  0.931 0.01548 0.00444 2.440e+005
Cp = 0.425073,   Ct = 0.643836
Power = 1.802796e+003 Watts, Thrust = 2.730605e+002 Newtons
```

Enter TSR: end with -ve: 10

Radius	iter.	aoa	a	Cl	Cd	deltor	Re
0.1300	13	3.85	0.227	0.735	0.01287	0.00074	2.726e+005
0.1900	6	4.09	0.243	0.753	0.01276	0.00115	2.914e+005
0.2500	5	4.62	0.268	0.793	0.01301	0.00157	3.005e+005
0.3100	7	5.09	0.288	0.830	0.01334	0.00197	3.032e+005
0.3700	8	5.37	0.298	0.852	0.01365	0.00235	3.030e+005
0.4300	8	5.47	0.299	0.860	0.01376	0.00270	3.025e+005
0.4900	8	5.44	0.297	0.858	0.01373	0.00303	3.030e+005
0.5500	8	5.38	0.296	0.853	0.01364	0.00336	3.046e+005
0.6100	9	5.33	0.297	0.849	0.01357	0.00367	3.069e+005
0.6700	10	5.31	0.298	0.847	0.01353	0.00397	3.088e+005
0.7300	9	5.28	0.298	0.845	0.01349	0.00424	3.095e+005
0.7900	10	5.28	0.295	0.844	0.01350	0.00447	3.084e+005
0.8500	11	5.29	0.291	0.846	0.01354	0.00462	3.057e+005
0.9100	12	5.24	0.285	0.842	0.01351	0.00455	3.032e+005
0.9700	13	4.85	0.262	0.812	0.01314	0.00372	3.041e+005

Cp = 0.461121, Ct = 0.811886

Power = 1.955683e+003 Watts, Thrust = 3.443329e+002 Newtons

Enter TSR: end with -ve: 12

Radius	iter.	aoa	a	Cl	Cd	deltor	Re
0.1300	33	1.26	0.196	0.498	0.01235	0.00058	3.097e+005
0.1900	10	1.52	0.224	0.518	0.01211	0.00092	3.388e+005
0.2500	4	2.10	0.273	0.573	0.01208	0.00128	3.534e+005
0.3100	6	2.64	0.318	0.622	0.01212	0.00161	3.587e+005
0.3700	9	3.00	0.351	0.655	0.01214	0.00189	3.598e+005
0.4300	10	3.22	0.370	0.675	0.01222	0.00214	3.601e+005
0.4900	10	3.33	0.380	0.685	0.01226	0.00238	3.612e+005
0.5500	11	3.39	0.390	0.690	0.01229	0.00259	3.636e+005
0.6100	12	3.45	0.401	0.696	0.01233	0.00279	3.666e+005
0.6700	13	3.53	0.414	0.703	0.01237	0.00295	3.692e+005
0.7300	13	3.60	0.422	0.710	0.01240	0.00311	3.702e+005

```
0.7900 13   3.69  0.427  0.718  0.01243  0.00324 3.690e+005
0.8500 14   3.82  0.428  0.729  0.01246  0.00335 3.660e+005
0.9100 15   3.94  0.420  0.739  0.01248  0.00332 3.631e+005
0.9700 17   3.81  0.373  0.728  0.01244  0.00260 3.643e+005

Cp = 0.417002,  Ct = 0.960945
Power = 1.768568e+003 Watts, Thrust = 4.075511e+002 Newtons
Enter TSR: end with -ve: -9
Enter the wind speed in m/sec: end with -ve: -9
>>
```

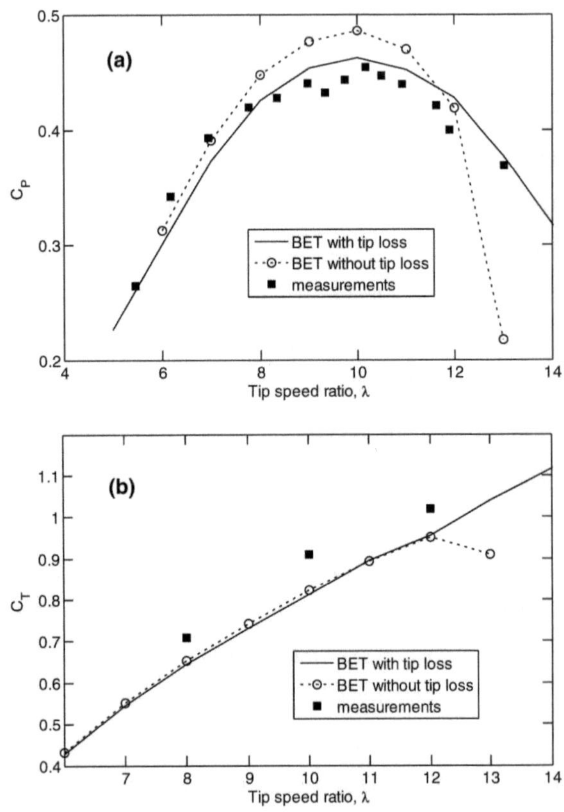

Fig. 5.2 **a** Predicted power coefficient as a function of tip speed ratio compared to the measurements of Anderson et al. [5]. **b** Predicted thrust coefficient as a function of tip speed ratio compared to the measurements of Anderson et al. [5]

Judging from Fig. 5.2a, the blade element predictions with tip losses predict the power to within about 10% and are most accurate around the point of optimum efficiency. The tip loss correction reduces the power, Fig. 5.2a, but makes little difference to the thrust, Fig. 5.2b. This under-prediction may not be general: the calculations by Shen et al. [1] for a different wind turbine over-estimated the

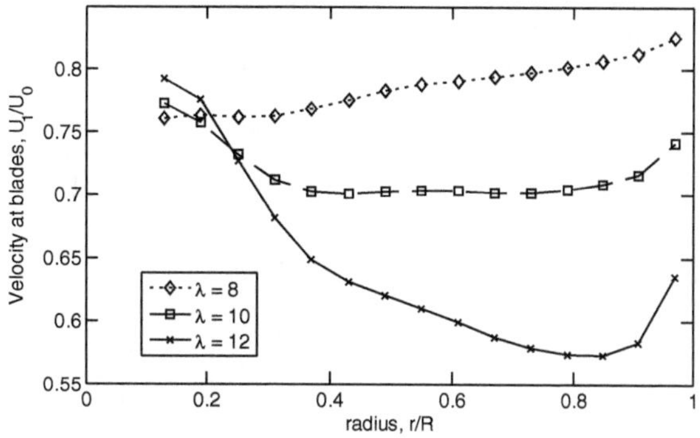

Fig. 5.3 Predicted velocity through the rotor for the measurements of Anderson et al. [5]

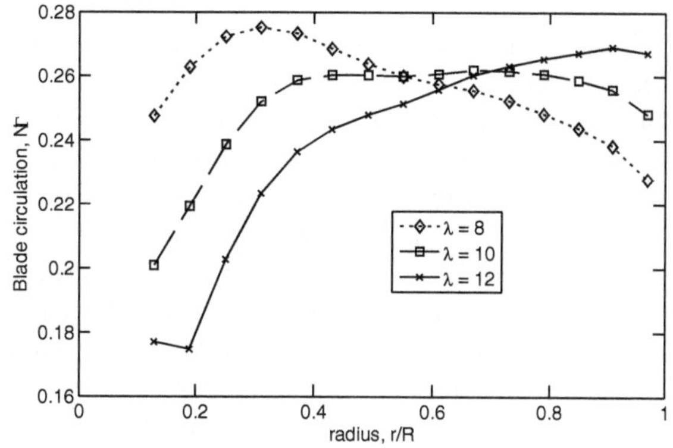

Fig. 5.4 Predicted blade circulation for the measurements of Anderson et al. [5]

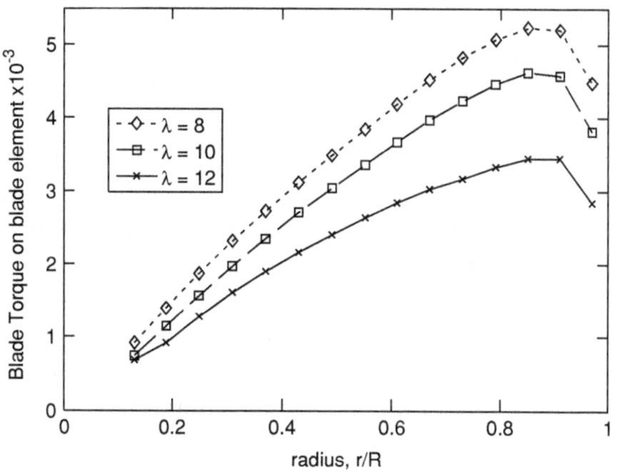

Fig. 5.5 Predicted distribution of blade torque for the measurements of Anderson et al. [5]

thrust. Unfortunately the maximum measured value of C_T in Fig. 5.2b is just over unity, so it was not possible to explore the accuracy of the calculations in the high thrust region; there just does not appear to be any good data for these conditions.

Fig. 5.6 Predicted distribution of blade thrust for the measurements of Anderson et al. [5]

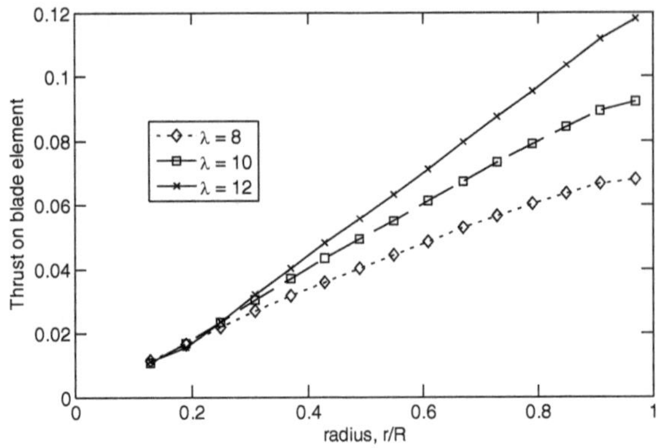

Having shown the reasonable accuracy of the blade element calculations for the rotor properties, it is necessary to consider the radial dependence. At the maximum efficiency point, $\lambda = 10$, a and hence U_1 is roughly constant with a value not much greater than the Betz–Joukowsky value of 2/3, Fig. 5.3. Note that the figure shows $U_1 = 1 - a$, not the value of the wind speed at the blades which would be $1 - a_{\hat{b}}$. The decrease in U_1 near the tips is a consequence of the tip loss correction; without this correction the distribution for $\lambda = 10$ would be almost flat for $r/R > 0.4$. Measurements of the velocity immediately behind the blades by Anderson et al. [5] for $\lambda = 10$ (only) show considerable scatter, but are in general agreement with the calculated values of U_1. Similarly, Γ is approximately constant over the blade, $\lambda = 10$ suggesting that the region where the bound vorticity decreases and the trailing vorticity is formed, is narrow. Γ is distributed quite differently on an efficient wind turbine than on a helicopter rotor in hover, but there are strong similarities with the behaviour of propeller blades. Finally, there is remarkably little variation in the Reynolds number along the blade for all three operating conditions. This is partly due to the reduction in chord as radius and effective velocity increase.

Figure 5.4 shows Γ is most uniform along the blade when λ is at its optimum value, and tends to decrease with increasing λ. The decrease in circulation near the tips is due to the tip loss correction. We will see later that, without this correction, $\Gamma\lambda$ is approximately constant for a wind turbine.

Figures 5.5 and 5.6 show that most of the power and thrust is produced near the tip, simply because of the rapidly increasing contribution of Ωr to U_T. It follows that the blade design near the hub is not critical for power extraction, so that modifications to accommodate the attachment to the hub and structural considerations, such as increasing the thickness of the blade to withstand the centrifugal loads, can be made without compromising power performance. We will see in Chap. 6 that starting performance introduces important aerodynamic considerations to the hub region which may work against these modifications.

Comparison of the power_calc.m output for blade element α with Fig. 5.1 suggests that, as expected, optimum performance occurs when l/d is maximised.

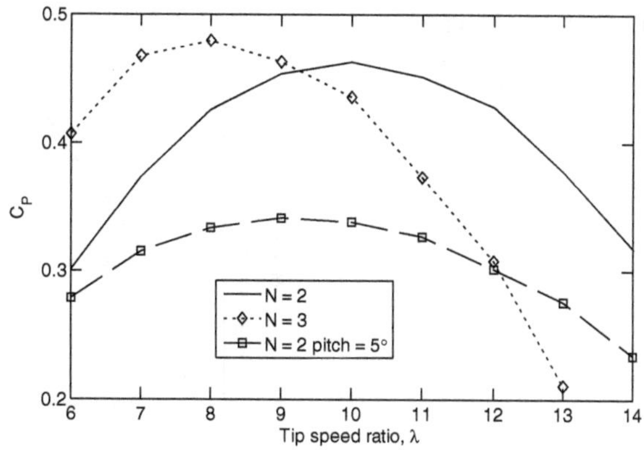

Fig. 5.7 Predicted effect of changing the number of blades and pitch on the power coefficient for the measurements of Anderson et al. [5]

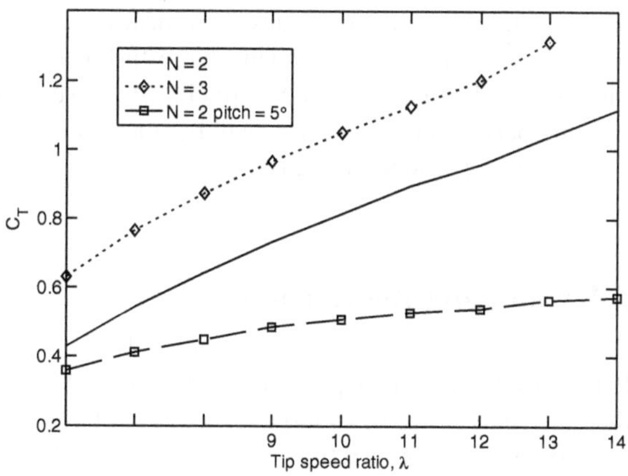

Fig. 5.8 Predicted effect of changing the number of blades and pitch on the thrust coefficient for the measurements of Anderson et al. [5]

In turn, this suggests that a turbine using a more modern blade section would be more efficient. Before discussing this issue, however, it is worthwhile to consider the effects of changing the number of blades, and altering the blade pitch angle without using a different aerofoil.

Figure 5.7 shows the two-bladed results from Fig. 5.2 compared to $N = 3$ along with the calculations for a two-bladed rotor with a pitch of 5°; the whole blade has been rotated such that θ_P in Fig. 3.2 is increased by 5° for all elements. Thus a typical chord line has moved away from the plane of rotation. The corresponding thrust data are plotted in Fig. 5.8. It is typical that increasing N does not have a large effect on the thrust and power levels; the main change is to the value of λ at which they occur. This is not surprising given that the blade chord is absorbed into the solidity, σ, in the blade element equations in Chap. 3. Thus it is the product Nc, rather than c that is important. In other words, apart from the effects of changing the Re, three blades with a certain chord will perform very similarly to two blades with the chord increased by 50%. As shown by the very large reductions in power and thrust, a pitch change of 5° has a marked impact on

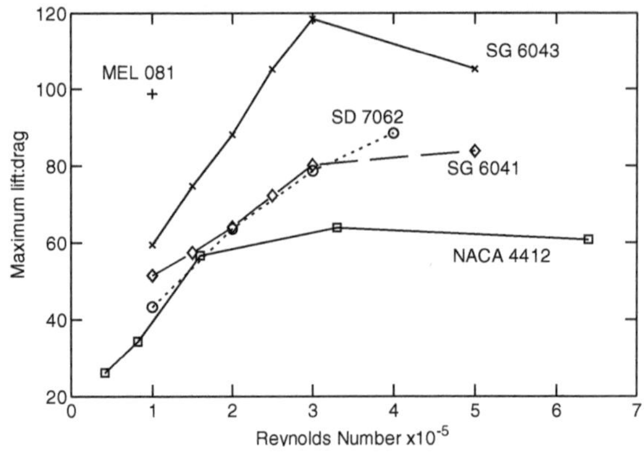

Fig. 5.9 Maximum lift:drag ratio for some small wind turbine aerofoils

performance. This demonstrates the efficacy of using pitch adjustment to control power output, as occurs on many large turbines. On the other hand, the C_P distribution is flatter at the higher pitch, which suggests that low wind speed and starting performance is improved by increasing the pitch. As we will see in Chap. 6, this is indeed the case.

5.4 Changing the Aerofoil

It has been mentioned previously that the NACA 4412 is an old aerofoil, and therefore may not give as large a power output as the designer would like because its maximum lift:drag ratio is lower than for more modern sections, as shown in Fig. 5.9. The SG6043 and the MEL 081 are particularly impressive, but unfortunately, the co-ordinates for the latter are not freely available. Of those sections plotted in Fig. 5.9, the SD7062, Lyon et al. [8], has been used extensively by the Wind Energy Group at Newcastle University. Its main advantage over the SG sections described in Chap. 4 is its increased thickness, nearly 14%, which gives it extra strength to withstand the high centrifugal loads on small blades. The available lift:drag data for this aerofoil are shown in Fig. 5.10. Calculations for the chord and twist distribution of the blade of Anderson et al. [5] are now described. The changes to power_calc.m to use this, or any new, aerofoil are trivial. They are:

```
%ReadIn_4412 % Read in Cl and Cd for the NACA 4412 section
ReadIn_7062 % Read in Cl and Cd for the SD7062 section
```

immediately after the chord and twist distribution are established and, of course, a data file must be prepared for the lift and drag of the new section.

For brevity, ReadIn_7062.m is not listed here, but it will be easily understandable for any reader who has mastered the corresponding functions for the

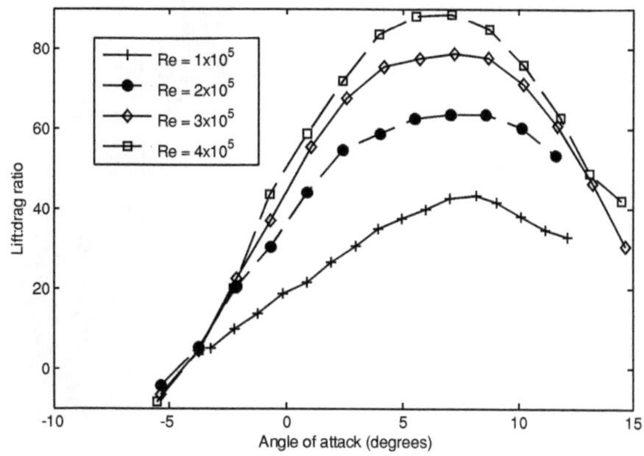

Fig. 5.10 Lift:drag ratio for the SD7062 aerofoil

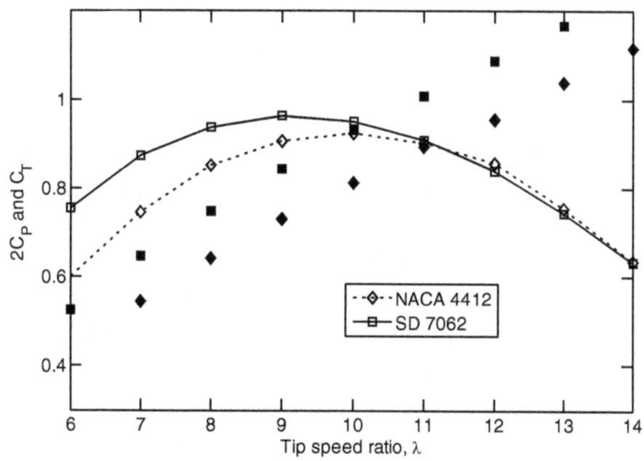

Fig. 5.11 Predicted effect of changing the aerofoil on the power and thrust coefficients for the measurements of Anderson et al. [5]. *Open symbols* show twice C_P, *closed symbols* show C_T

NACA 4412. The computed power and thrust coefficients are compared in Fig. 5.11 to the computed results for the NAC 4412 section from Fig. 5.2. As expected, there is a modest improvement in C_P that reflects the increase in the lift:drag ratio for the SD7062 compared to the NACA 4412.

5.5 Maximising Power Extraction

In preparation for the more complete study in Chap. 7, the optimising of blade performance is now considered in terms of maximising C_P. Recall from Chap. 2 that the Betz–Joukowsky limit requires $C_P \leq 16/27 = 0.593$ for unshrouded turbines, and from Chap. 1, that the high-*Re* blade of the large Vestas V47 has $C_P \approx 0.5$ (when allowance is made for mechanical and electrical inefficiencies). Furthermore, it is known that a finite N, and a finite l/d, reduce the maximum C_P from the Betz–Joukowsky limit. Thus it would appear that a maximum C_P of just

Fig. 5.12 Visualization of the tip vortex of a two-blade 10 m diameter wind turbine in the NASA Ames 80 by 120 ft wind tunnel. Photo from Dr Scott Schreck, NREL

above 0.50 is a reasonable target with the current design methodology. This suggestion is reinforced by the study by Selig and Coverstone-Carroll [9]; they used "genetic algorithms" to search for optimal blade designs using aerofoil sections including those considered in this chapter and in Chap. 3. They obtained a maximum C_P of 0.53. "Evolutionary" optimisation for multi-dimensional blade design is the subject of Chap. 7.

In addition to a high C_P, it is important to have as broad a range of λ as possible over which performance is close to the optimal. The need for this was noted for constant speed turbines in the discussion of Figs. 1.5 and 1.6. It is also true for variable speed turbines, even though the reasons are less obvious. A narrow or "peaky" performance curve would require a very accurate control system to adjust the blade speed to maintain the optimum λ as the wind speed varies. This task is made more difficult because the wind speed is not usually monitored on small turbines, whereas nearly all large machines have an anemometer mounted on the nacelle. Furthermore, some wind speed and direction changes occur too rapidly for a turbine to follow without suffering huge dynamic loads, so that even variable speed turbines of necessity operate at varying λ.

There are a number of ways in which blade performance can be related to the Betz–Joukowsky limit. The first is just to compare the actual C_P to the Betz–Joukowsky value of 16/27 and attempt some sort of trial-and-error searching. Here the comparison is taken further by the apparently circuitous route of investigating the structure of the wake behind the blades in terms of its vortex structure that was introduced in Chap. 3. By Kelvin's theorem for the conservation of circulation in an otherwise inviscid fluid, the simplest possible consequence of the generation of bound vorticity by the blades is that vorticity of strength Γ must be shed into the wake at the blade tips, much as the bound vorticity of an aircraft's wing feeds the wing tip vortices that are often visualized in contrails. In contrast, blade tip vortices are helical, as demonstrated by the beautiful flow visualization of Fig. 5.12. In the far-wake—well downstream of the blades—R_∞, the wake and vortex radius, and p_∞, the pitch, which measures the distance between successive turns of the

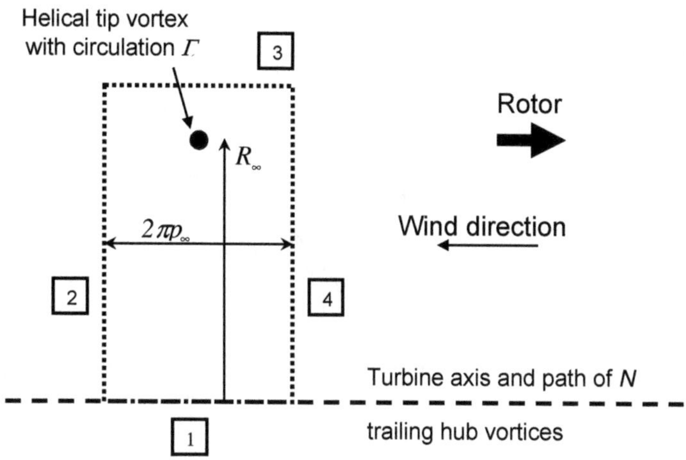

Fig. 5.13 Rectangular contour in the far-wake that encloses one tip vortex

helix, are constant. Furthermore the *N* hub vortices can be assumed to lie along the axis of rotation, so that they make no contribution to U_∞, although they do determine W_∞. This wake vortex geometry was designated the "Joukowsky model" by Okulov and Sørensen [10] in honour of the famous Russian aerodynamicist who first conceived it. If the vortices are assumed to be vortex lines, that is, they have no thickness, it is straightforward to determine the relationship between Γ, p_∞, and U_∞, for the Joukowsky wake. Since $U_\infty = 1/3$ at the Betz-Joukowsky limit, an expression can be derived for the corresponding Γ.

The first step involves a circulation argument much like that used in Sect. 4.5. Consider a far-wake consisting of only one helical tip vortex, with pitch p_∞ and radius R_∞, as indicated schematically in Fig. 5.13 which shows a slice through the wake so the view is from the side of Fig. 5.12. Only the vortex within the dotted contour is displayed and the wind speed is ignored for the present (Fig. 5.13).

The first leg of the contour of length $2\pi p_\infty$, is along the rotor axis; $2\pi p_\infty$ is also the distance between points on a helical vortex separated by one revolution (2π). The radial second and fourth legs (of equal length) are longer than R_∞. Thus the integral of the vorticity over the area enclosed by the circuit (which contains just one vortex) is Γ which must equal the circulation around the circuit. The contribution from the two radial legs will cancel as they are $2\pi p_\infty$ apart, and if the radial legs are much larger than R_∞, the contribution from the third leg will be zero. Thus the circulation is $2\pi p_\infty U_\infty$ where U_∞ is the velocity along the rotor axis. To complete the argument, reduce the radial legs so that they are smaller than R_∞. The circulation contained by the contour is now zero and this shows that U_∞ is independent of the radius in the far-wake, as is assumed in the one-dimensional analysis that leads to the Betz-Joukowsky limit. In general, there will be *N* tip vortices contained within the circuit and a non-zero wind speed, so the far-wake velocity is

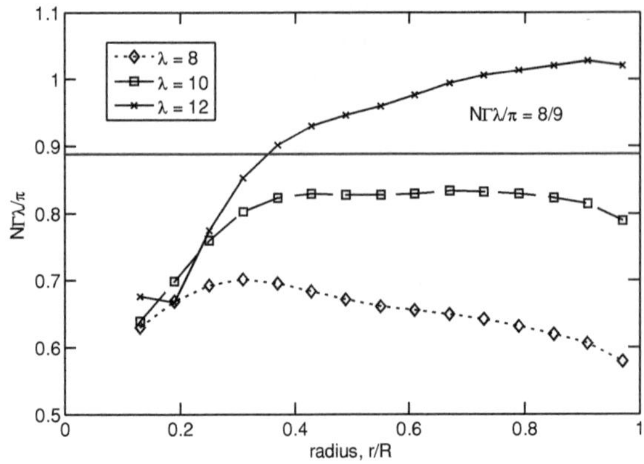

Fig. 5.14 Normalised circulation for $\lambda = 8$, 10, and 12

$$U_\infty = 1 - \frac{N\Gamma}{2\pi p_\infty} \tag{5.8}$$

where the second term on the right is subtracted for consistency with the convention that the circulation is positive. Now assume that the velocity of the tip vortices in the direction of the wind is the average of the wake velocity and the wind speed. Thus

$$p_\infty = \frac{1 + U_\infty}{2\lambda} \tag{5.9}$$

which seems entirely reasonable, but is hard to prove, see [11, 12], and is only approximately true at high λ. From (5.8) and (5.9),

$$U_\infty = \sqrt{1 - \frac{N\Gamma\lambda}{\pi}} \tag{5.10}$$

At the Betz–Joukowsky limit, the circulation is given by

$$N\Gamma\lambda/\pi = 8/9 \tag{5.11}$$

It is worth noting that (5.11) can be derived from the different starting point of conservation of angular momentum, see Eq. 5.44 of Spera [6] and Exercise 5.2.

For a given N, (5.11) gives the target Γ to achieve maximum performance for any λ. For comparison with the data in Fig. 5.4, for example, Eq. 5.11 in combination with (4.12b) gives $N\Gamma = 0.348$, 0.278, and 0.232 near the tip for $\lambda = 8$, 10, and 12 respectively. By comparing these values to those shown in Fig. 5.4 redrawn in Fig. 5.14, it is clear that the best performance occurs at $\lambda = 10$ for the NACA 4412 section.

If (5.11) is combined with Eq. 4.12b, it is easy to show that

$$cC_l \approx \frac{16\pi}{9N\lambda^2 r} = \frac{16\pi}{9N\lambda\lambda_r} \tag{5.12a}$$

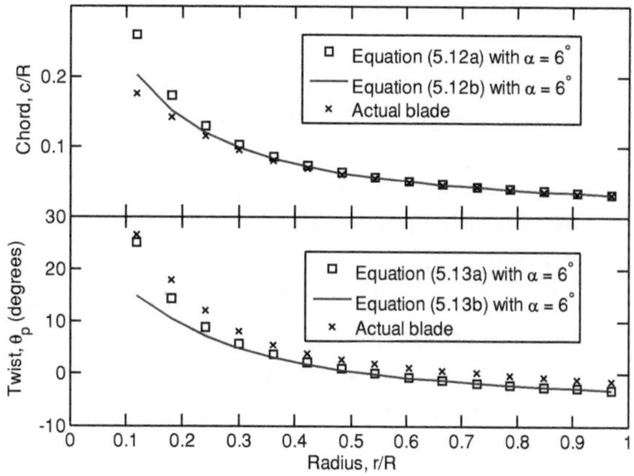

Fig. 5.15 Optimum chord and twist compared to Anderson et al. [5] blade

C_l, is usually taken as $C_{l,\max}$, the value at which l/d is maximised since drag has been ignored in obtaining (4.12b). Equation 5.11 immediately gives the chord at each radius once the number of blades and the tip speed ratio have been chosen. Equation 5.12a agrees (at high λ) with the more general, and more difficult to derive, "optimal chord" Eq. 3.67a of Burton et al. [13].

$$cC_l = \frac{16\pi}{9N\lambda\sqrt{4/9 + [\lambda_r + 2/(9\lambda_r)]^2}} \tag{5.12b}$$

Ignoring a' on the grounds that it is small, Eq. 3.12 becomes

$$\tan\phi \approx 2/(3\lambda_r) \tag{5.13a}$$

which determines ϕ. Knowing the angle at which the chosen C_l occurs, then gives θ_P by Eq. 3.8. Equation 5.13a is a high-λ approximation to Eq. 3.68a of Burton et al. [13]:

$$\tan\phi = \frac{2}{3\lambda_r + 2/\lambda_r} \tag{5.13b}$$

Once the designer has selected the aerofoil profile for the blade, the tip speed ratio and tip radius, Eqs. 5.12 and 5.13 can be used to complete the blade aerodynamic design.

An example of the use of Eqs. 5.12 and 5.13 is shown in Fig. 5.15 which compares the equations with the output from tcdist.m with the NACA 4412 section, $U_0 = 10$ m/sec, and $\lambda = 10$. As shown above, power_calc.m gave the best performance at $\lambda = 10$ with C_P over 0.46. The tip Re is just over 3×10^5. Figure 5.1 indicates that $C_{l,\max}$ will occur at $\alpha = 6°$ and the data (not shown but can be viewed in n4412.in) gives $C_{l,\max} \approx 0.90$. The chord and twist of the actual blade closely follow the theoretical optimum curves, with a discrepancy of about

1° in twist. If the blade twist is reduced by this amount, the C_P increases slightly to 0.465 at $\lambda = 10$.

Of course the combination of Eqs. 5.12a and 5.13a will not be adequate at small r, but Fig. 5.5 shows that the hub region contributes little to the overall power production. As explained by Burton et al. [13] this fact is exploited by the designers of large blades to improve their manufacturability. Most have linearly tapering chord which approximates the $1/r$ optimal shape for the outer blade but is more manageable near the hub. However, Chap. 6 shows that the hub region is crucial for good starting and low wind speed performance, implying that linear taper should not be used for small blades. Chapter 7 shows that the hub region of small blades can be designed largely for starting in the context of optimising *both* starting and power extraction.

The discussion of optimum efficiency has not considered the second order effects of tip losses. These can be included but the details are complex, [3], and so will not be addressed here. The numerical optimisation described in Chap. 7 includes tip losses.

5.5.1 Further Reading

The NREL web site has comprehensive documentation on their wind turbine activities: http://www.nrel.gov/wind/. The so-called "Unsteady Aerodynamics Experiment", during which the photo in Fig. 5.12 was taken, is described at: http://wind2.nrel.gov/amestest/

5.5.2 Exercises

1. Derive Eq. 5.7.
2. Show that Eq. 5.10 can be derived by combining Eqs. 2.9, 2.15, and 2.16.
3. A common question from interested non-specialists in wind turbine theory, is "Well, if you can get x kW power from a turbine with two blades, why cannot you get $1.5x$ kW by adding another blade?" How would you answer that question?
4. If the contribution to the thrust from each blade element, ΔC_T, is linear in r, show that the radial point of action of the thrust occurs at $r/R = 2/3$. Comment on the significance of this result for the root bending moment of the blade.
5. If the ΔC_p is linear in r, what is the point of action of the torque?
6. For a turbine operating at the Betz–Joukowsky limit, determine the approximate magnitude of W_∞ as a function of radius, r.
7. The tip vortex in Fig. 5.12 does not appear to increase in diameter. What does that say about the operating condition at which the photo was taken?

8. Visit the UIUC web site (URL given in Sect. 4.6) and download the lift and drag data for one of the SG6040, SG6041, or SG6043. Modify the programs used in this chapter to read in the new data and determine the effect on performance of changing the aerofoil section. Do not change the blade twist and chord from that given by tcdist.m.
9. Using tcdist.m to vary the pitch, and then run power-calc.m with the SD7062 lift and drag, estimate the maximum possible C_P and compare with the C_P from the optimal twist and chord distributions given by Eqs. 5.12 and 5.13.
10. For the chord distribution of Eq. 5.12a, show that the Reynolds number, Re, is approximately constant along a blade operating optimally at a high tip speed ratio.
11. For the Joukowski model of the blades and wake, explain why only the helical tip vortices induce flow through the blades.

References

1. Shen WZ, Mikkelsen RF, Sørensen JN, Bak C (2005) Tip loss corrections for wind turbine computations. Wind Energy 8:457–475
2. Xudong W, Shen WZ, Wei JZ, Sørensen JN, Ch J (2009) Shape optimisation of wind turbine blades. Wind Energy. doi:10.1002/we335
3. Clifton-Smith M (2009) Wind turbine blade optimisation with tip loss correction. Wind Eng 33:477–496
4. de Vries O (1979) Fluid dynamic aspects of wind energy conversion. Technical report, AD-A-076315, Advisory Group for Aerospace Research and Development, 92-Neuilly-sur-Seine (France)
5. Anderson MB, Milborrow DJ, Ross JN (1982) Performance and wake measurements on a 3 m diameter horizontal axis wind turbine. In: 4th international syrupon wind energy systems, pp 113–135
6. Spera D (1994) Wind turbine technology. ASME Publishing, New York
7. Miley SJ (1982) A catalog of low Reynolds number airfoil data for wind turbine applications. Report prepared under U.S. DoE Wind Energy Program
8. Lyon CA, Broeren AP, Giguere P, Gopalarathnam A, Selig MS (1997) Summary of low-speed airfoil data: 3. Soartech
9. Selig MS, Coverstone-Carroll VL (1996) Application of a genetic algorithm to wind turbine design. J Solar Energy Eng 118:22–28
10. Okulov VL, Sørensen JN (2007) Stability of helical tip vortices in a rotor far wake. J Fluid Mech 576:1–25
11. Okulov VL, Sørensen JN (2010) Maximum efficiency of wind turbine rotors using Joukowsky and Betz approaches. J Fluid Mech 649:497–508
12. Wood DH, Boersma J (2001) On the motion of multiple helical vortices. J Fluid Mech 447:149–171
13. Burton T, Sharpe D, Jenkins N, Bossanyi E (2001) Wind energy handbook. Wiley, Chichester

Chapter 6
Starting and Low Wind Speed Performance

6.1 Introduction

It will be shown in this and the next chapter that designing blades only for efficient power extraction using Eqs. 5.12a, b and 5.13a, b for the optimum chord and twist respectively, can result in a high cut-in wind speed. The turbine will then be of little practical use except in rare situations where the wind blows regularly and strongly. Even a turbine like the 1.94 m diameter one shown in Fig. 6.1, whose power curve, Fig. 6.2, demonstrates a good cut-in wind speed of 3.5 m/s, may not extract maximum power in low winds.

There are at least three reasons to be particularly interested in performance at low wind speed. First, many small turbines are located close to the load they supply and this may not be a good wind site. Secondly, very few small turbines have pitch adjustment, so that a blade designed for optimal power extraction at a high tip speed ratio, will present high angles of attack when it is stationary. As shown in Fig. 4.9 these high angles occur at low Reynolds numbers which makes it even more difficult to generate sufficient lift to turn the blades. Thirdly, turbines start only when the aerodynamic torque generated on the stationary blades exceeds the resistive torque in the generator and drive train. Recall that Fig. 1.12 shows typical resistive torques of PMGs used in small turbines as a fraction of rated torque. For comparison, Sect. 6.2 describes an estimation of the ratio of aerodynamic starting to rated torque and shows that it can be smaller than the resistive torque ratio. In the absence of significant Reynolds number effects, the aerodynamic torque scales as R^3—as shown by Eq. 2.9 Figure 1.12 suggests that the resistive torques decrease less rapidly than R^3 as R decreases; in fact they increase as a proportion of rated torque. Thus starting is a particularly important issue for micro-turbines. This is most likely the reason why micro-turbines for power on yachts, for example the Rutland 913 in Fig. 1.3, often have five or more blades.

The turbine in Fig. 6.1 had a permanent magnet generator (PMG) with a maximum static resistive torque of 0.36 Nm. It was mounted for many years on the roof of the Engineering Building at Newcastle University and has been investigated in

D. Wood, *Small Wind Turbines*, Green Energy and Technology,
DOI: 10.1007/978-1-84996-175-2_6, © Springer-Verlag London Limited 2011

Fig. 6.1 500 W 1.94 m diameter turbine with cup anemometer and wind vane

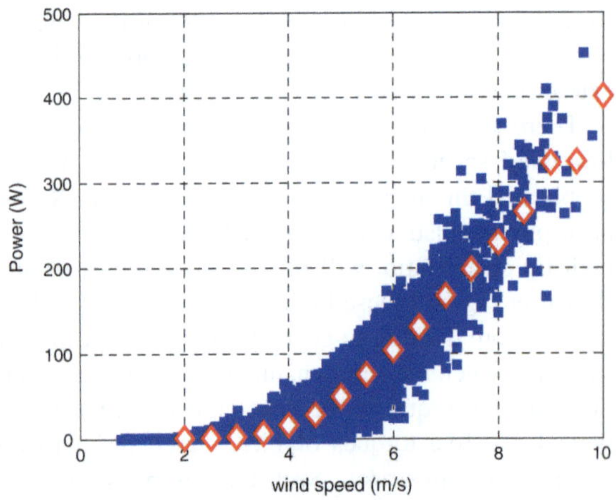

Fig. 6.2 500 W turbine power curve. 1-min averages of power output (*solid blue squares*) have been placed in wind speed bins of 0.5 m/s width, and averaged to give *open red diamonds* from Wright [1]

detail. All the starting measurements described in this chapter come from that turbine. Its power curve, Fig. 6.2, suggest that it had good low wind performance in that it has a low cut-in wind speed, defined in IEC 61400-2 as the "lowest mean wind speed at hub height at which the wind turbine produces power". Figure 6.3, however, graphically demonstrates that the average starting wind speed is significantly higher than the cut-in speed. The rest of this chapter analyses rotor acceleration from rest under the action of the aerodynamic torque caused by the wind. It will be assumed that the turbine maintains perfect alignment with the wind during starting. In practice, however, the combination of long starting times, as shown in Fig. 6.4,

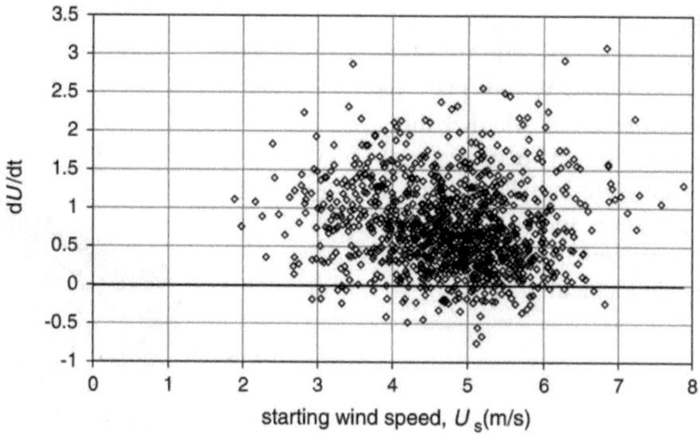

Fig. 6.3 Measured starting wind speeds for the 500 W turbine from Wright [1]. Average starting wind speed = 4.8 m/s. The units of dU/dt are m^2/s

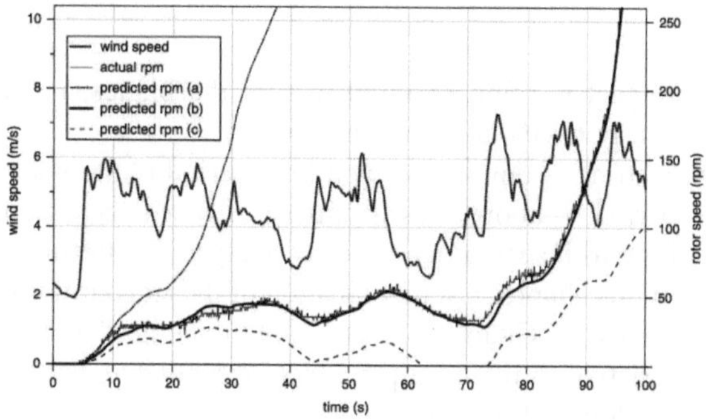

Fig. 6.4 Low speed start of the 500 W turbine, taken from Wright [1]

Table 6.1 Parameters for turbine in Fig. 6.1 from Wright [1]

Parameter	Value
Hub radius	0.25 m
Inertia of composite blades and hub	0.41 kg m^2
Inertia of generator	0.02 kg m^2
Total rotational inertia	0.43 kg m^2
Dynamic resistive torque	0.24 Nm
Static resistive torque	0.36 Nm

and the propensity of wind direction changes to increase with decreasing wind speed, means that most starting sequences will be affected by yaw errors. Yaw is likely to reduce the starting torque and should lead to the over-prediction of starting behaviour.

Figure 6.4 shows a typical starting sequences of the 500 W turbine, one of nearly 200 sequences analysed by Wright [1]. Table 6.1 lists the main turbine parameters related to starting: note that the blade holder made a negligible contribution to the rotor inertia. The difference between the static and dynamic resistive torque is likely to be caused by friction in the bearings as cogging torque should be independent of generator speed. Figure 6.4 shows a number of important features of starting:

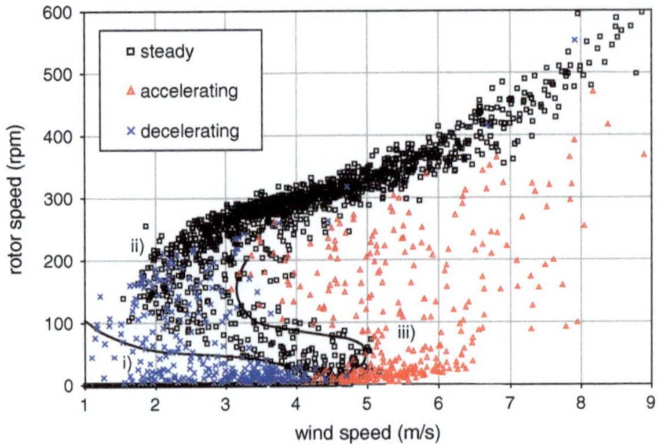

Fig. 6.5 Low wind speed performance of 500 W wind turbine from Wright [1]. The *symbols* define whether the rotor is steady, accelerating, or decelerating. The *solid lines* show the boundaries of the regions as determined by blade element analysis

- A gust is often required to turn the blades, in this case from about 2 m/s to nearly 6 m/s around time, $t = 3$ s.
- Starting is dominated by a long period of slow acceleration—called the "idling period"—about 85 s in this case. It will be assumed that idling is sufficiently long to justify a quasi-steady analysis of starting. The analysis to be developed in this chapter is a modification of the standard, power-producing, blade element theory of Chap. 5. Note from Fig. 1.12 that this turbine has a high ratio of resistive to rated torque which accentuates, but is not uniquely responsible for, the long idling period.
- Using the high-α formulations of aerofoil lift and drag considered in Chap. 4, it is possible to accurately simulate starting as shown by curve (b) in the figure. This curve, along with the less accurate predictions labelled (a) and (c) will be explained in Sect. 6.4.

Wright [1] measured about 200 h of low wind speed performance of this turbine. Using the method described by Wright and Wood [2] he determined whether the rotor was steady, accelerating or decelerating for a large number of 15 s sequences taken from the data. Figure 6.5 shows the results in terms of rotor speed versus wind speed with the rotor state denoted by symbol. The delineation between the three regions—(1) decelerating, (2) steady, and (3) accelerating as determined by blade element theory is shown by the solid lines. It is clear that the rotor accelerated from rest only for wind speeds of approximately 4.8 m/s, the "starting" wind speed from Fig. 6.3. However, the decelerating rotor did not stop until about 2.5 m/s, which can be called the "stopping" wind speed. It is clear that the cut-in wind speed from Fig. 6.2 is an average of the starting and stopping speeds and that the key to improving low wind speed performance is to understand starting performance. The former task begins the next section which discusses the aerodynamic torque on a stationary blade for comparison with the resistive torque data from Chap. 1. Improving starting performance is part of the multi-dimensional optimisation described in Chap. 7.

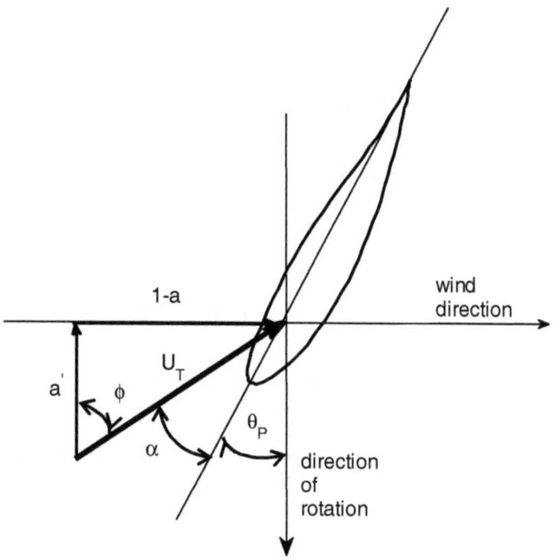

Fig. 6.6 Blade element at radius r on a stationary blade

6.2 Estimating the Starting Torque

Figure 6.6 shows the velocity diagram for a stationary blade element which can be compared to that for a rotating, power producing element in Fig. 3.2. Both a and a' are likely to be small on a stationary blade, so the velocity vector is not to scale. It is known from the previous two chapters that typical twist angles are around 20° near the hub and zero near the tip, so the angles of attack on a stationary blade are high and in the range discussed in Sect. 4.4. Thus the aerodynamic torque on a stationary blade is likely to be small and the resistive torque, as documented in Fig. 1.12, often significant.

The term "rotational" inflow factor for a' is not appropriate for a stationary blade so it will be referred to as the "circumferential" inflow factor. Equation 3.11 for the torque generated on each blade element can be written as

$$\frac{dQ}{dr} = \frac{1}{2} N \rho U_T^2 c (C_l \sin \phi - C_d \cos \phi) r \tag{6.1}$$

Assuming that both a and a' are small:

$$\phi \approx \pi/2, \quad \alpha \approx \pi/2 - \theta_P, \quad \text{and} \quad \sin \alpha \approx \cos \theta_P \tag{6.2}$$

In words: only the lift generates torque on a stationary blade. In order to determine typical values of the starting torque, Q_s, Eq. 4.6 with $A = 1$ gives, after some easy manipulation,

$$\frac{dQ_s}{dr} = \frac{1}{2} N \rho U_T^2 cr \sin(2\theta_\rho) \tag{6.3}$$

Balancing (6.3) against the angular momentum in the wake, from Eq. (3.5), leads to

$$(1-a)WU = \frac{1}{2}\sigma U_T^2 \sin(2\theta_p) \tag{6.4a}$$

When the blade is stationary, there is no need to distinguish between U_0 and U_1 so U is used for the wind speed. (This equivalence of wind speeds will also be assumed for the starting analysis.) Similarly, W is the constant (with downstream distance) circumferential velocity behind the blades. The axial momentum equation gives

$$a(1-a)U^2 = \frac{1}{2}\sigma U_T^2 \cos^2\theta_p \tag{6.4b}$$

it follows from (6.4a) and (6.4b) that

$$a \approx \frac{1}{2}\sigma \cos^2\theta_p \tag{6.5a}$$

and

$$a' = a \tan\theta_p \tag{6.5b}$$

Typically, $\sigma < 0.2$ for modern blades, and θ_P lies in the range 0–30°, so that $a' < a$ as well as a, $a' \ll 1$ so the assumption made about the wind speed after (6.4a) is justified. Further, $U_t \approx (1-a)U \approx U$. Alternatively, $U_t/U \approx 1$ and a and a' can (and will) be neglected in the analysis of starting torque. By normalising all lengths by R, and all velocities by U, Q_s (in Nm), is determined by

$$Q_s = \frac{1}{2}N\rho U^2 R^3 I_{cp} \tag{6.6}$$

where the "chord-pitch integral" I_{cp}, and its integrand, i_{cp}, are defined by

$$I_{cp} = \int_{r_h}^{1} i_{cp}dr = \int_{r_h}^{1} cr \sin(2\theta_p)dr \tag{6.7}$$

The lower limit on the integral, the "hub" radius r_h, is assumed to be the beginning of the aerodynamic section of the blade. It is straightforward to integrate (6.7) for an optimal power-producing blade because cr is constant by Eq. 5.12a:

$$cr = \frac{16\pi}{9N\lambda_p^2 C_{l,\max}} \tag{6.8}$$

where λ_p is the design tip speed ratio for rated power. Note that the constancy of cr means that the integral in (6.7) receives its largest contributions where θ_P is largest, that is, in the hub region. Thus most starting torque is generated near the hub.

With $\alpha_{\max}$ again defining the angle for maximum lift:drag, it follows from (5.13a) that

$$\tan\,\theta_p = \frac{2 - 3\lambda_p r\,\tan\,\alpha_{\max}}{3\lambda_p r + 2\,\tan\,\alpha_{\max}} \tag{6.9a}$$

so

$$\sin\,\theta_p\cos\,\theta_p = -\frac{1}{2}\sin\,(2\alpha_{\max}) + \frac{4\,\sin(2\alpha_{\max})}{4 + 9\lambda_p^2 r^2} + \frac{6\lambda_p r\,\cos(2\alpha_{\max})}{4 + 9\lambda_p^2 r^2} \tag{6.9b}$$

where, interestingly, the first term depends only on the choice of aerofoil, and the last two on the design tip speed ratio as well. Integration of (6.7) using Eq. 6.8 for cr, gives the expression for the coefficient of the stationary blade torque, Q_s:

$$C_{Q,s} = \frac{Q_s}{\frac{1}{2}\rho U_s^2 \pi R^3} = \frac{16}{9\lambda_p^2 C_{l,\max}}(I_1 + I_2 + I_3) \tag{6.10}$$

where U_s is the starting wind speed. The integrals on the right are, in the same order as the integrands in (6.9b):

$$I_1 = -(1 - r_h)\sin(2\alpha_{\max}) \tag{6.11a}$$

$$I_2 = \frac{4\sin(2\alpha_{\max})}{3\lambda_p}\left[\tan^{-1}\left(\frac{3\lambda_p}{2}\right) - \tan^{-1}\left(\frac{3\lambda_p r_h}{2}\right)\right] \tag{6.11b}$$

and

$$I_3 = \frac{2\cos(2\alpha_{\max})}{3\lambda_p}\ln\left(\frac{4 + 9\lambda_p^2}{4 + 9\lambda_p^2 r_h^2}\right) \tag{6.11c}$$

To estimate the rated torque, assume that the turbine operates at the Betz-Joukowsky limit at rated wind speed U_p. It is also assumed that the generator and drive train have no inefficiencies. Thus

$$C_{Q,p} = \frac{Q_p}{\frac{1}{2}\rho U_p^2 \pi R^3} = \frac{16}{27\lambda_p} \tag{6.12}$$

Figure 6.7 shows the ratio of the stationary torque from (6.10) to the rated torque from (6.12) for the parameter values given in the caption. Recall from Fig. 1.12 that the ratio of resistive to rated generator torque typically falls between 0.01 and 0.02, then even the highest curve in Fig. 6.7 should give cause for concern.

When it is further realised that (6.8) allows c to become infinite as $r \downarrow 0$, and a practical limit on blade chord is likely to be around 0.1 then the situation gets worse. This is demonstrated by the calculations for $c_{\max} = 0.1$ in the figure with the twist distribution unaltered. These calculations are somewhat more involved than for the unconstrained chord and are not described. Further, N now becomes important, and Fig. 6.7 shows that three blades produce significantly more torque than two. In summary: Fig. 6.7 shows how much attention the blade designer has

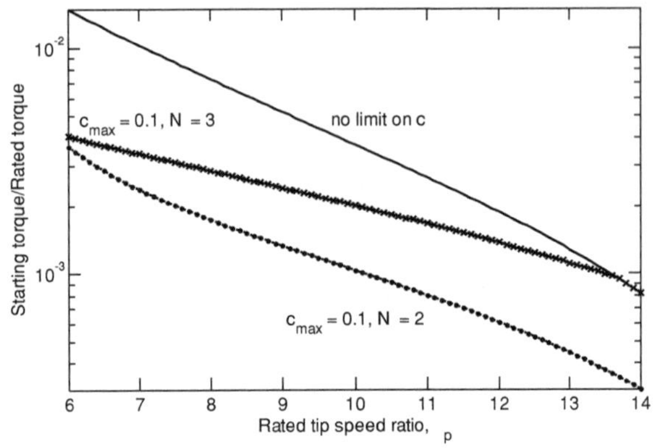

Fig. 6.7 Ratio of starting to rated torque. $U_s/U_p = 1/3$, $C_{l,\max} = 1, \alpha_{\max} = 5.7°$. The vertical axis is logarithmic

to pay to starting behaviour even when the turbine has a low resistive torque. Furthermore, the reduction in torque when the blade chord is limited again shows the significance of the hub region in generating the starting torque.

6.3 Analysis of Starting

Obviously a stationary blade cannot extract power, which is a consequence of neglecting a and a', but an accelerating one must at least extract its own rotational kinetic energy. However, it is tempting to generalise the analysis of Sect. 6.2 for the whole starting sequence by assuming that no power is extracted, a and a' remain negligible, and the aerodynamic torque acts solely to accelerate the rotor. These assumptions will remain valid for small values of the ratio of the rotor kinetic energy at the end of the idling period, $1/2NJ\Omega^2$ to the kinetic energy of the wind that passed through the rotor during time T_s, $1/2\rho U_s^3 \pi R^2 T_s$. For the sequence in Fig. 6.4, with $NJ = 0.43$ kg m^2 from Table 6.1, $\Omega = 75$ rpm, $U_s = 4$ m/s, and $T_s = 80$ s, the ratio is 0.0014. Figure 6.8 shows that T_s scales with U_s^{-2}so that eventually, the assumption of minimal power extraction must break down. However, the designer's aim is to achieve starting at the lowest sensible U_s so the assumption is retained.

Unsteadiness can alter the lift and drag of the blade elements because the magnitude of the velocity and the angle of attack are both changing with time. To estimate their effects, consider only the latter, which can be gauged from the "reduced frequency", $k = \pi\dot{\alpha}c/(2U)$where $\dot{\alpha} = d\alpha/dt$ is the rate of change of the angle of attack measured in rad/s, see, for example, Sect. 8.4 of Leishman [3]. For a rotating blade, U_T is

$$U_T = \sqrt{U^2 + (r\Omega)^2} = U\sqrt{1 + \lambda_r^2} \tag{6.13}$$

From the geometry of Fig. 6.6,

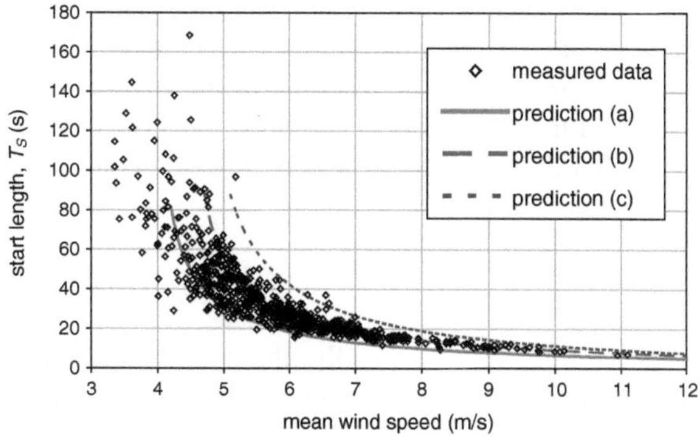

Fig. 6.8 Starting time and the mean wind speed of 665 measured starting sequences, and the predicted curves using the three simulation methods assuming steady *U* [1]. The predictions (a), (b), and (c) are by the methods used in Fig. 6.4

$$\tan\phi \approx \frac{U}{\Omega r} = \frac{1}{\lambda r} = \frac{1}{\lambda_r} \tag{6.14}$$

so that

$$\dot{\alpha} = \frac{d\alpha}{dt} = \frac{-rU}{r^2\Omega^2 + U^2}\frac{d\Omega}{dt} \tag{6.15}$$

With the data used to assess the importance of the rotor kinetic energy, and $c = 0.128$ at the hub, $r = 0.25$ m, and $c = 0.043$ m at the tip, $r = 0.97$ m, $k \approx 8.5 \times 10^{-5}$ at the tip and 8.4×10^{-5} at the hub. These values are too low to cause the lift and drag to deviate from quasi-steady values [3].

Equations 4.6 and 4.7 with $A = B = C = 1$ provide the easiest treatment of starting and are the only ones used here. Wright [1] found that these equations had to be modified for the turbine in Fig. 6.1 with its relatively low aspect ratio of about 9.0. Recent measurements of the starting of 2.5 m long blades with $AR = 14.3$ were best reproduced with the high lift equations to be used here as shown at the end of this chapter. Equation 6.3 becomes

$$\frac{dQ}{dr} = N\rho U^2\left(1 + \lambda_r^2\right)^{1/2} cr \sin\theta_p\left(\cos\theta_p - \lambda_r \sin\theta_p\right) \tag{6.16}$$

after some elementary manipulation to remove α and ϕ in favour of θ_p.

When all lengths are normalised by the blade tip radius, *R*, and all velocities by *U*, (6.16) the equation for *Q*, the aerodynamic torque acting on the starting rotor, is

$$Q = N\rho U^2 R^3 \int_{r_h}^{1} \left(1 + \lambda_r^2\right)^{1/2} cr \sin\theta_p\left(\cos\theta_p - \lambda_r \sin\theta_p\right) dr \tag{6.17}$$

Equation 6.17 is probably too messy to integrate analytically even when *cr* is constant. Nevertheless, it has several interesting consequences. First, the starting time, say the time taken for the rotor to accelerate from rest to a specified tip speed ratio, should be linear in the number of blades as long as solidity effects do not alter the blade element lift and drag. Secondly, the starting time should scale as U^{-2}, as is demonstrated in Fig. 6.8. Thirdly, it is easy to show that *Q* has a local maximum when $\lambda = 0$ because the drag does not influence Q when Ω and λ are zero. Q then *reduces* as the rotor accelerates from rest and the drag reduces the torque. The initially high torque in the hub region decreases, and, generally, the torque in the outer part of the blade increases as the blade accelerates. However, this is not the whole story. For (6.17) with constant cr, dQ/dr goes through zero at the radius where $\theta_p = 0$:

$$r \approx \frac{2}{3\lambda_p \tan \alpha_{\max}} \tag{6.18}$$

Since $\alpha_{\max} \approx 6°$ typically, the most efficient blade will have regions of negative starting torque whenever $\lambda_p \geq 7$ approximately, which is the case for many small wind turbines. If the more general assumption is made that C_l and C_d are determined by the aerofoil's pressure distribution at high α, so that $C_l/C_d = 1/\tan \alpha$ and using the result of Exercise 6.2, it can be shown that dQ/dr goes through zero when

$$\lambda_r = 1/\tan \alpha \tag{6.19}$$

Positive blade element torque again requires positive θ_p. Both (6.18) and (6.19) are approximate, and are to be taken only as indications that there may be regions of negative aerodynamic torque on the outer part of starting blades. More aerofoil data and measurements of starting rotors are required to resolve this issue. Finally, the R^3 dependence indicates the difficulty in starting small rotors.

The integral in Eq. 6.17 is easily evaluated in blade element form, which corresponds to using the midpoint rule for the quadrature.

An important consequence of the assumption that no power is extracted during starting is that the rotor torque Q acts only to accelerate the blades. Thus at any time during starting

$$\frac{d\lambda}{dt} = \frac{R(Q - Q_r)}{JU} \tag{6.20}$$

where J is the total rotational inertia and Q_r is the resistive torque. For most wind turbines, J is dominated by the contribution from the blades as shown by the discussion of Fig. 1.14 and Table 6.1 for the 500 W turbine. When $Q_r = 0$, N cancels when (6.17) is equated to (6.20). Thus: *starting is independent of the number of blades in the absence of resistive torque*, unless the local solidity of the blade is high enough to alter the lift and drag. The next section shows how Eq. 6.20 can be evaluated using standard methods for solving ordinary differential equations (ODEs).

The predictions using (6.17) are labelled as (c) in Figs. 6.4 and 6.8 and are not as accurate as predictions (b) which were obtained by accounting for low aspect ratio effects at high incidence as described by Clifton-Smith et al. [4]. However, as noted earlier, (6.17) was found to be more accurate for the newer measurements on the higher aspect ratio 2.5 m long blades.

Current knowledge of high-α aerodynamics at low Re is too rudimentary to provide any firm conclusions on the appropriate formulation of lift and drag. As discussed in Chap. 4, the issues of aspect ratio are not understood and its effects have been inconsistently formulated in blade element theories. The situation is complicated by the implication from Fig. 4.6b that what appear to be aspect ratio effects may, in fact, be due to *Re* differences between the 2.5 and 0.97 m blades. More fundamental experimental information is needed. In the meantime Eq. 6.16 will be used in the expectation that the dual optimisation to be pursued in the next chapter requires only the relative starting performance of the competing blades.

6.4 Estimating the Rotor Inertia

Equation 6.20 shows that the rotor inertia is needed to calculate starting. For simplicity only blades with a uniform density, ρ_b, are considered and the blade attachment is ignored. The moment of inertia, *J*, of a rotor of *N* blades about the *x*-axis (the turbine axis) is given by

$$J = N\rho_b \int (y^2 + z^2)dxdydz \tag{6.21}$$

where the *z*-axis is in the radial direction and *y* is in the direction of rotation of the blade. The integration is over the blade volume. (The normalisation by blade radius is delayed until Eq. 6.25.) The blade-fixed Cartesian co-ordinate system used in (6.21) has its origin on the axis of rotation with *y* in the direction of the wind, and *z* along the blade. Because BET does not constrain the position of the blade elements along their chord, the determination of *J* at the design stage always carries some uncertainty. It is assumed that the centroids of the elements lie along the *z*-axis, and the *z*-position of a blade element is its radius, *r*. Equation 6.21 can be rewritten as

$$J/(N\rho_b) = \int r^2 dxdydr + \int y^2 dxdydr = J_1 + J_2 \tag{6.22}$$

where J_1 should dominate as $c/r \ll 1$ for most wind turbines. The first integral is just

$$J_1 = A \int (cr)^2 dr \tag{6.23}$$

where *A* is the dimensionless result of dividing the area of the aerofoil section by the square of the chord. As determined by trapezoidal integration, the values of *A* for several aerofoils are listed in Table 6.2.

An approximate expression for J_2 can be found by assuming each blade section is rectangular with thickness t' such that $A = t'/c$, and the centroid is along the *z*-axis. Now change to x',y' co-ordinates such that the latter is along the chord line. Thus

$$J_2 \approx \int_0^R dr \int_{-c/2}^{c/2} dy' \int_{-t'/2}^{t'/2} (\cos^2\theta_p y'^2 + \sin^2\theta_p x'^2 + 2\cos\theta_p \sin\theta_p x'y')dx' \tag{6.24a}$$

Table 6.2 Values of A for several aerofoil sections

Aerofoil	A
SD7062	0.08818
SG6040	0.10411
SG6041	0.06955
SG6042	0.06916
SG6043	0.06850
NACA0012	0.08213
NACA4412	0.08211

$$= \frac{t'c^3 \cos^2\theta_p + t'^3 c \sin^2\theta_p}{12} \tag{6.24b}$$

When all lengths are normalised by the blade tip radius, R,

$$J = N\rho_b A R^5 \left[\int (cr)^2 dr + \frac{A}{12}\left(\int c^4 \cos^2\theta_p dr + A^2 \int c^4 \sin^2\theta_p dr \right) \right] \tag{6.25}$$

where the first integral in the square brackets corresponds to J_1 and the second to J_2. The relative magnitude of the integrals is at most $(c/r)^2$. The rotor shown in Fig. 6.1 had composite blades but subsequently these were replaced by timber ones for which $\rho_b = 550$ kg/m^3 and $J = 0.190$ kg m^2 from Eq. 6.25 when the rectangular attachment section was included. This value agrees with that obtained from finite element analysis. The ratio of the second to first term in (6.25) was 3.23×10^{-3}, which justifies the approximate treatment of J_2. To look at the magnitude of the terms another way: for a three bladed turbine with $\lambda_p = 7.5$, and $C_{l,\max} = 1$, $cr \sim 0.03$. If further, $A \sim 0.1$, and $r_h \sim 0.1$, the three integrands are 9×10^{-4}, 6.75×10^{-5}, and 6.75×10^{-7} at the hub. Thus the contribution to J_2 is an order of magnitude less than that to J_1 even when the chord at the hub is 30% of the tip radius.

6.5 Matlab Program for Starting

This section describes a Matlab program start_calc.m which reads in the blade characteristics in exactly the same manner as power_calc.m (described in Chap. 5), calculates blade inertia, and then determines the time to start for user-specified wind speed in two ways. First, Matlab's adaptive Runge–Kutta 4th/5th order solver ode45 is used to solve Eq. 6.20 with no resistive torque. This solution is expected to be very accurate. The second calculation uses the standard Adams–Moulton method, designated ABM4 in Sect. 12.2.2 of Fausett [5], which has only two function evaluations per step and is, therefore, more attractive for the large number of starting calculations needed for numerical design optimisation in the next chapter. The program listing is:

```
function start_calc(Numb, U_start, rho_blade, lambda_start,delt)
% Find the time to start of a rotor using Matlab Runge-Kutta routine
% ode45 and the Adams-Moulton method
% This version assumes no resistive torque
.
% Read in rad_ch_tw.dat as in power_calc.m  (code not shown)
.
% Determine c1, c2 in the equation for d(lambda)/dt
c1 = 1.2*Numb*U_start^2*r_tip^3*delr;
% Find J
Area = 0.06850; % Area of SG6043 aerofoil
A2 = Area^2;
J1 = sum((chord.*rad).^2);  % Eq. 6.23
J2 = sum((cosd(twist).^2) .* (chord.^4)) + A2*sum((sind(twist).^2) .* (chord.^4));
% Convert to dimensioned inertia
J = Numb*delr*rho_blade*Area*r_tip^5*(J1 + J2/12);  % Eq. 6.25
c2 = r_tip/(J*U_start);
tmax=5;
[t, lambda]=ode45(@(t, lambda) deriv( t, lambda,c1, c2,chord,...
                          twist, rad),[0 tmax],0.0);
istart=find(lambda>lambda_start,1);% Find first value > lambda_start
time_to_startRK=t(istart1)+(t(istart)-t(istart-1))*(lambda_start-...
                      lambda(istart-1))/(lambda(istart)-lambda(istart-1));
fprintf(' Time to start from RK = %5.3f seconds\n', time_to_startRK)
isize=tmax/delt;
lambdaAM=zeros(isize,1); timeAM=zeros(isize,1);
Fn_3=0.0; Fn_2=0.0; Fn_1=0.0; % Initial values of derivative
i=1;
for i=1: isize % Use Adams Moulton for integration
   Fn_0 = deriv(timeAM(i), lambdaAM(i), c1, c2, chord, twist, rad);
   lam_pred=lambdaAM(i) +delt*(55*Fn_0-59*Fn_1+37*Fn_2- 9*Fn_3)/24;
   Fp_1 = deriv(timeAM(i), lam_pred, c1, c2, chord, twist, rad);
   del_l = delt*(9*Fp_1+19*Fn_0-5*Fn_1+Fn_2)/24;
   if (lambdaAM(i) + del_l > lambda_start)
             time_to_startAM = timeAM(i) + delt*(lambda_start - lambdaAM(i))/del_l;
             fprintf(' Time to start from AM = %5.3f seconds\n' time_to_startAM)
             break
   end
   lambdaAM(i+1) = lambdaAM(i) + del_l;
   timeAM(i+1) = timeAM(i) + delt  ;
   Fn_3=Fn_2;
   Fn_2=Fn_1;
   Fn_1=Fn_0;
end
end %start_calc

function dlamdt = deriv(t, lambda, c1,c2,chord, twist, rad)
lamr = lambda.*rad;
lamr2=lamr.*lamr;
tmp = sqrt(1 + lamr2).*sind(twist).*(cosd(twist) - lamr.*sind(twist)).* chord.*rad;
dlamdt=c1*c2*sum(tmp);  % Combine Eq. 6.17 and 6.20
end % deriv
```

The arrays t and lambda hold the time and tip speed ratio respectively as calculated by the Matlab function:

```
[t, lambda]= ode45(@(t,lambda)deriv(t, lambda,c1,c2,chord,...
                                    twist, rad),[0 tmax], 0.0);
```

Note the syntax for invoking ode45 and passing the extra (but necessary) information to the function deriv to calculate the derivative. ode45, along with all Matlab ODE solvers, requires the dependent and independent variables to be passed to the derivative function as arrays. In Matlab, operations that are to be carried out term-by-term are preceded by a period (.), of which there are four in the following line:

```
tmp = sqrt(1 + lamr2). * sind(twist). * (cosd(twist)
      -lamr. * sind(twist)). * ...chord. * rad;
```

Matlab functions, such as sqrt and sind always operate term-by-term. tmp is then summed to find the total aerodynamic torque on each blade. The treatment of tmp is an example of vectorisation, an alternative to operations that would normally be done in a loop in other languages. Looping however, is very time consuming in Matlab and is to be avoided if possible. Issues of computational time become very important in Chap. 7 in discussing numerical optimisation that requires many thousand evaluations of objective functions.

It is possible to make ode45 stop at λ = lambda_start, but this is somewhat cumbersome to program. It is easier to ensure that tmax, the upper limit on the ODE solution, is larger than the time to start, then use the Matlab function find to locate the first value of t for which λ exceeds the final value for starting and then interpolate linearly to find the time to start.

Below is a snippet of a Matlab session showing a run of the program for a three-bladed turbine with $R = 1.5$ m with the same chord and twist distribution as used by Anderson et al. [6], and made from a material with density 550 kg m^{-3}. The time step, δt, for the Adams–Moulton method is 0.1 s. The starting wind speed is 5 m/s and starting is deemed to be completed when $\lambda = 1$.

```
>> start_calc(3,5,550,1,0.1)
Time to start from RK = 4.022 seconds
Time to start from AM = 4.010 seconds
>>
```

Figure 6.9 shows the calculated starts with the adaptive Runge–Kutta method as the solid line and the Adams–Moulton solutions with varying delt have the symbol indicated. The integration is not difficult and it is to be expected that accurate results can be obtained, especially for comparative purposes, from relatively large values of delt, even though start_calc.m does not provide accurate starting values for the Adams–Moulton method for δt.

Fig. 6.9 Calculated starting sequence using adaptive Runge–Kutta (RK) and Adams–Moulton method (AM) for δt indicated

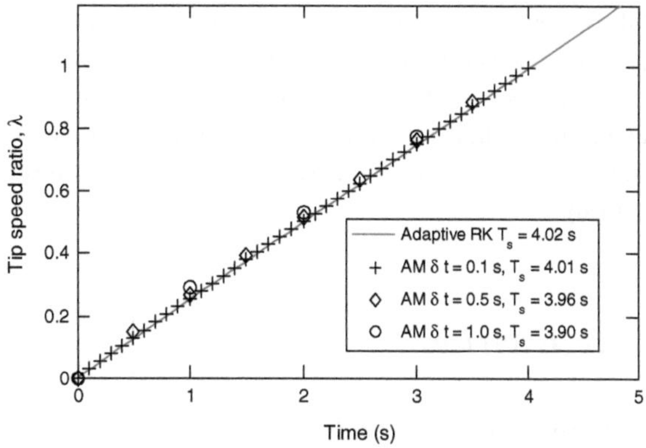

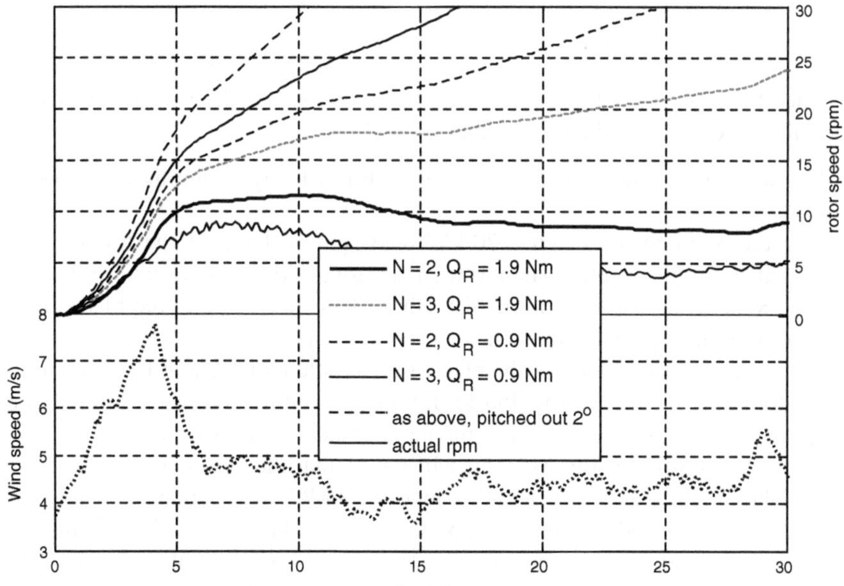

Fig. 6.10 A failed measured starting sequence of the Aerogenesis 5 kW turbine with two blades and an induction generator and gearbox

It is not difficult to extend start_calc.m to cope with variable wind speeds as was done for Fig. 6.4 and for Fig. 6.10 which shows a failed starting sequence for the Aerogenesis 5 kW turbine shown in Fig. 1.3 with a gearbox resistive torque of 1.9 Nm. The thick solid line shows that the calculated starting performance overestimates the measured Ω, possibly because of yaw effects which were ignored in the calculation. It can be seen that using three blades would have allowed the turbine to start, indicating that the resistive torque is significant; otherwise starting performance would have been independent of N. After these measurements were taken, new gearboxes were sourced which had a resistive torque of 0.9 Nm. The calculations show that the turbine would then start with two blades and that using three blades would again improve starting but not by as large an amount. However, using three blades would allow them to be pitched out by 2° and still produce the same power as two blades. With this pitch, three blades provide even better starting.

6.5.1 Exercises

1. Derive Eqs. 6.9a and 6.9b.
2. Show that for any form of C_l and C_d, Eq. 3.11 can be written as

$$\frac{dQ}{dr} = \frac{1}{2} N\rho U^2 \sqrt{1 + \lambda_r^2} cr(C_l - \lambda_r C_d)$$

when there is no axial or circumferential inflow. Hence derive Eq. 6.16 for the special case of $C_l = 2 \sin \alpha \cos \alpha$ and $C_d = 2 \sin^2 \alpha$.

3. Use the above result for dQ/dr to derive Eq. 6.19 by assuming only that $C_l/C_d = 1/\tan \alpha$.
4. You are required to design a 20 kW blade on the basis of the 5 kW design described in this chapter. The turbine is to use a PM generator whose designer can guarantee a cogging torque of 8 Nm or less. In designing the blade would you make any special provision for the starting performance?
5. The twist and chord data for the blade used to obtain the data in Figs. 6.2 to 6.4 is stored in the file baby_blade.dat in the online materials http://extras.springer.com. Use that data to estimate the blade's aspect ratio.
6. Looking carefully at Fig. 6.1 you will notice that the aerodynamic section of the blades starts at a larger radius than the PMG cover, mainly because the blade was designed before the importance of the hub region for starting was appreciated and was designed to be mounted in the existing blade holder. How would you redesign the blade holder (and possibly the blade attachment) to improve starting?
7. Assume it is possible to obtain a PMG for the 500 W turbine whose power curve is shown in Fig. 6.2, with a cogging torque of 0.2 Nm and that the new generator would cause the cost of the turbine to increase by 15%. Estimate the new cut-in wind speed and the amount of extra power that would be extracted given the Rayleigh distribution of wind speed with a mean wind speed of 5.7 m/sec., as statistics described in Sect. 1.6. Could you justify using the new generator?
8. You are told to design blades for a small wind turbine using Eqs. 5.12a and 5.13a for a given generator with no resistive torque, U_p, λ_p, R, and aerofoil section. The only choice you have is whether $N = 2$, 3, or 4. Which value

would you use to minimise the starting time? What effects will the choice of N have on the efficiency of power extraction?

9. Run the program start_calc.m for a value of tmax much greater than the starting time. What happens?
10. Modify the program start_calc.m to use other Matlab Runge–Kutta solvers. How does the choice of solver influence the starting time? You should find that there is very little change. Why?
11. Alter start_calc.m to include a resistive torque. Run the calculation for the parameters in Sect. 6.5 but with $Q_r = 0.25$ Nm. How does the starting time change? What effect does resistive torque have on the scaling of T_s with U?

References

1. Wright AK (2005) Aspects of the aerodynamics and operation of a small horizontal axis wind turbine. PhD thesis, School of Engineering, University of Newcastle
2. Wright AD, Wood DH (2004) The starting and low wind speed behaviour of a small horizontal axis wind turbine. J Wind Eng Ind Aerodyn 92:1265–1279
3. Leishman JG (2006) Principles of helicopter aerodynamics, 2nd edn. CUP, Cambridge
4. Clifton-Smith M, Wood DH, Wright AD (2010) Optimising design and construction for wind energy systems in low wind environments. In: Sørensen JD, Sørensen JN (eds) Wind energy systems: optimising design and construction for safe and reliable operation. Woodhead Publ, Cambridge, pp 370–391
5. Fausett LV (2008) Applied numerical analysis using Matlab. Pearson Prentice Hall, New York
6. Anderson MB, Milborrow DJ, Ross JN (1982) Performance and wake measurements on a 3 m diameter horizontal axis wind turbine. In: 4th international syrupon wind energy systems, pp 113–135

[illegible]

Chapter 7
Blade Design, Manufacture, and Testing

7.1 Introduction

Most small turbines do not have pitch adjustment of the blades whereas most large turbines do. The reasons are mainly associated with cost and complexity, but they make starting performance a subject worthy of a whole chapter. The development of blade design in this chapter assumes no pitch adjustment. There are three main results from Chaps. 4–6 that must be kept in mind in considering blade design. First, power extraction occurs mainly on the outer part of any turbine blade. Large blades usually have a linear taper approximating the theoretical optimum $1/r$ dependence of the chord near the tip. The deviations closer to the hub are of little consequence. Second, starting torque is generated mainly near the hub. Thirdly, thick aerofoil sections are not to be used for small turbine blades as their low-Re performance is generally very poor at high angles of attack. The first two results are encouraging as they suggest that a reasonable compromise between high efficiency and good starting is possible. Section 7.3 shows that to be the case, in the context of a numerical optimisation method described in the next section. The third result is cause for concern because thick sections are used near the hub of large blades for structural strength (in combination with the circular root attachment that allows pitch adjustment). It is for these reasons that the biggest difference between the shape of optimal blades for large and small turbines will occur in the hub region.

7.2 Optimisation Method

Designing a wind turbine blade for the single criterion of maximum power extraction is straightforward: once the operational tip speed ratio and aerofoil section(s) have been chosen, Eqs. 5.12 and 5.13 fix the chord and twist[1] at least

[1] Note that the large differences between (5.12a, b) and (5.13a, b) occur near the hub where little power is produced.

D. Wood, *Small Wind Turbines*, Green Energy and Technology,
DOI: 10.1007/978-1-84996-175-2_7,

when tip losses are ignored. However, maximum efficiency is only one of the parameters to be optimised on many blades. For example, large blades must produce minimal noise, and as discussed in the last chapter, small blades should usually start quickly in low winds. It is generally not possible to derive analytic expression analogous to (5.12) and (5.13) for multi-dimensional optimisation. Nevertheless, there are a vast number of multi-dimensional optimisation methods available for blade design, a number of which are available as Matlab functions. The strategy described here is not necessarily the best, but it (a) is the one the author is most familiar with, (b) reproduces known optimum results such as Eqs. 5.12 and 5.13 as special cases, (c) is very easy to code, (d) is easily extended to higher dimensional optimisation, and (e) is computationally very fast.

This chapter considers optimisation methods using an "evolutionary strategy", [1]. These seek to mimic the process of natural selection to arrive at an "optimum" solution by "evolving" a population over a sufficient number of generations. They start with a randomly-generated initial population, breed new members, determine the fitness of existing and new members, and then decide which members live and which ones die. The breeding must introduce the equivalent of mutation to the "genes" of each new member. In the present case, the genes are the twist and chord of each blade element as it will be assumed that the same aerofoil profile is used for the whole blade.

The particular method is called "differential evolution", DE. In its basic form, DE generates a new population by the following process. For each member or vector of the current generation, $\mathbf{x}_i$, say, a comparison vector, $\mathbf{c}_i$, is constructed from a basis vector $\mathbf{b}_i$, or from a trial vector $\mathbf{t}_i$ according to

$$\mathbf{t}_i = \mathbf{b}_i + w(\mathbf{u}_i - \mathbf{l}_i) \tag{7.1}$$

where w is a weighting factor and $\mathbf{b}_i$, $\mathbf{u}_i$ and $\mathbf{l}_i$ are randomly chosen members of the current population that are different from each other and from $\mathbf{x}_i$. Following [1], w is fixed at 0.8. A random number of genes from $\mathbf{t}_i$ is added to the complementing genes from $\mathbf{b}_i$ to form $\mathbf{c}_i$. Genes are chosen for $\mathbf{c}_i$ using a "crossover factor", CR, whose function is best described by the following pseudo-code:

```
FOR j = 1, number of genes
IF (rand < CR)
     ci(j) = ti(j)          ! Take gene from trial vector
ELSE
     ci(j) = bi(j)          ! Take gene from base vector
END DO
```

Following the recommendation of Price et al. [1] CR = 0.10. The fitness of $\mathbf{c}_i$ is determined as

$$\text{fitness}(\mathbf{c}_i) = a\frac{C_p(\mathbf{c}_i)}{\max(C_p)} + (1-a)\frac{\min(T_s)}{T_s(\mathbf{c}_i)} \tag{7.2}$$

where the two "objective functions" to be optimised—the power coefficient, C_P, and the inverse of the starting time, T_s—are calculated at user-input wind speeds. The factor a is also input by the user. Note that the form of (7.2) maximises the fitness when searching for minimal starting times. The maximum C_p and minimum T_s are determined from the current population.

The fitness of $\mathbf{x}_i$ is also found from (7.2), and the blade ($\mathbf{x}_i$ or $\mathbf{c}_i$) with the higher fitness is retained for the next generation. To encourage continuing genetic diversity, no member of the population is allowed to survive more than 20 generations.

Obviously a controls the relative importance of efficiency and starting time in determining the fitness. It has been found from many calculations that $a = 1$ always gives the most efficient blade design in terms of closely approximating (5.12) and (5.13), and this blade is often the slowest to start. Both these features will be demonstrated in the example design considered later in this chapter. On the other hand there is no minimal starting time equivalent to the Betz-Joukowsky limit. A moment's thought will indicate the reason for this: for any R, the blade inertia depends on the distribution of $(cr)^2$ from (6.25) whereas Q depends on cr from (6.17). In the absence of Re effects, the smaller the chord the faster the blade will start. Thus the fastest blade has infinitely small chord and is not useful in practice. *All judgements of starting time must be relative.*

Provided the number of generations and the size of the population are appropriate, calculations for varying a should provide a good estimate of the "Pareto front" defined as the subset of blades for which at least one objective function is larger than that for every other blade. The members of the Pareto front, alternatively called the "optimal fitness front", are the "non-dominated" blades. An example Pareto front is given later. There is no single optimum blade for a two-dimensional optimisation, and the best that a blade designer can do is chose one blade on the Pareto front.

In the present application, it was found that assigning a single random chord and separate single random twist to each member of the initial population, gave the best results. In addition, the parameters listed in Table 7.1 are required as input by the user.

7.3 Matlab Programs for Optimisation

The objective functions in the programs to be described are modifications of the blade element methods for power extraction (Chap. 5) and starting time (Chap. 6). These programs are modified as described below but are not listed. All the programs and scripts are available from the online materials http://extras.springer.comand further description is given in the information file accompanying them. In particular there is a Matlab script file `blade_opt_setup.m` used to define the many important blade parameters which are held in the data structure `data`. The function that runs the optimisation is listed below:

```
function blade_deopt(data)
tic;    % record start time

creator; % run creator script to generate initial population
perc_in = 0.05; %The reporting interval (percentage)
percent = int8(double(data.no_gen)*perc_in);

fid = fopen(data.fitfile,'w'); % open output file for fittest blades
disp(' gen. Max. Cp for blade Min. Start Time for blade')

% calculate fittness of initial population
blade.fitness = survival(data, blade.chord,blade.twist);

% adjust fitness for poorly performing blades
index = find(blade.fitness(:,2) < data.Cp_min);
if (~isempty(index))
     blade.fitness(index,2) = 0;
end

% evolve blade population over no_gen generations
for i=1:data.no_gen
     % Find the fittest members of the current population
     [i_max_fit, max_fit, blade,data] = fittest(i,data,blade);
     data.max_fit(i,:) = max_fit;
     if (mod(i,percent)==0)   % If it is time to report ...
             fprintf('%5.1f%% complete,    non-dominated; ...
           %5.1f%%   Unfit; %5.1f%%\n',100.0*i/ data.no_gen,...
            100*data.no_nondom/data.no_indiv,...
             data.no_unfit/data.no_indiv*100)
             disp(' gen. Max. Cp for blade Min. Start Time for blade')
             for j=1:data.no_indiv
                         if (~blade.dom(j))
                                     fprintf(fid,'%12i %12.7f %12.7f
                            \n',j,blade.fitness(j,1),blade.fitness(j,2));
                         end
             end
     end
     if i< data.no_gen  % breed population, except for last gen
             [blade, data] = breed(i, i_max_fit, max_fit,data,blade);
   end
end
fclose(fid); % close file for fittest blades

index = find(blade.dom==0);
% Find the 'best' blade by averaging non-dominated blades
lindex=length(index);
if lindex >= 1
      fprintf('\n %3i blades are non-dominated \n \n', length(index))
      best_blade.chord = blade.chord(index,:);
      best_blade.twist = blade.twist(index,:);
elseif isempty(index)
      fprintf(' No blades are non-dominated \n')
end
% prepare to output average of non-dominated blades
best_blade.chord(lindex+1,:)=mean(best_blade.chord);
best_blade.twist(lindex+1,:)=mean(best_blade.twist);
for i = 1:lindex+1
 if (i <= lindex)
   fprintf(' For blade number %5i \n', index(i));
   iprint = 0;
 else
```

```
    fprintf(' For the average of the non-dominated blades \n');
    iprint=1;
end
best_blade.fitness(i,1)=power_calc_opt(data,best_blade.chord(i,:),
    best_blade.twist(i,:),iprint);
J = inertia(best_blade.chord(i,:),best_blade.twist(i,:),data);
delt = min(data.t_max/50, 0.25); % Time step for starting
best_blade.fitness(i,2)=start_calc_opt1(data,best_blade.chord(i,:),
    best_blade.twist(i,:),J, delt);
fprintf(' Cp = %5.3f, Starting time = %6.3f seconds \n',...
    best_blade.fitness(i,1), best_blade.fitness(i,2))
fprintf('\n   radius   chord   twist \n');
out=[data.rad' best_blade.chord(i,:)' best_blade.twist(i,:)'];
disp(out)
end
disp(' ');
fprintf('Computation time: %g seconds \n',toc);
end
```

The script creator generates the initial random population, and then references survival to compute the fitness of each member. Then the main loop is entered to repeat this process for the specified number of generations. It will be shown below that the members of the final population converge in C_P and T_s, so it makes sense to take the average of the non-dominated blades as the output of the optimisation. This averaging also reduces any random fluctuations in chord and twist. Note that the Matlab line continuation mark (…) has been removed from the two lines for the best_blade_fitness in the interests of brevity.

The script creator.m is listed below. The chord and twist of the initial population are chosen randomly but within the limits specified. The chord and twist are equal for all blade elements of each member. The arrays to hold the blade fitness, whether the blade is dominated, and its age are set up for later use.

Table 7.1 Input parameters for blade design for the Ginlong PMG 500 A permanent magnet generator

Parameter and value	Parameter and value
$N = 3$	$\rho_b = 550$ kgm^{-3}
$U_s = 5$ m/s	$U_p = 10$ m/s
$\lambda_f = 1.0$	$\lambda_p = 6.10$
$R = 1.06$ m	$\mathbf{r_h = 0.125}$
Maximum $c/R = 0.2$	Minimum $c/R = 0.01$
Maximum $\theta_p = 25°$	Minimum $\theta_p = -5°$
Maximum power = 754 W	**Cogging torque = 0 and 0.5 Nm**
Generator inertia = 0.006 kgm^2	Aerofoil section SG 6043

```
% Script creator to generate intial population
blade.chord = zeros(data.no_indiv,data.no_genes);
blade.twist = zeros(data.no_indiv,data.no_genes);

blade.fitness = zeros(data.no_indiv,2);
blade.dom = repmat(true,data.no_indiv,1);
blade.age = ones(data.no_indiv,1);

delc = data.max_chord - data.min_chord;
delt = data.max_twist - data.min_twist;

for i = 1:data.no_indiv   % Generate the first population
  rnd1 = rand;  % The blade chord and twist are random, but are
  rnd2 = rand;  %  lie between max and min values
  for j = 1:data.no_genes
     blade.chord(i,j) = rnd1*delc + data.min_chord;
     blade.twist(i,j) = rnd2*delt + data.min_twist;
  end
end
```

The function breed.m is referenced in the main loop. It contains all the code necessary to determine fitness—the array score holds the result of Eq. 7.2 for every member of the population. There are a number of features of the function that increase its execution speed. For example, Matlab allows contextual declaration of arrays but intial declaration of arrays, here using the Matlab function zeros, will result infaster program execution. Similarly it would be possible to implement the pseudo-code for determining the new genes in a number of loops, but Matlab's execution of loops is notoriously slow, so as many as possible of the operations are vectorised as was the determination of the derivative for the blade starting calculations in Chap. 6. Consistent with this approach is the determination of the indices for the basis vectors as a random permutation of the existing blades:

$$\mathsf{basis\_index} = \mathrm{randperm}(\mathsf{num\_indiv});$$

which makes every member of the current generation a "parent" of one and only one potential member of the next. An alternative would be to randomly choose the indices from the current population but that would require looping. Tests of the two alternatives did not show any significant differences in the final results so the simpler implementation was retained.

```
function [blade, data] = breed(iter, i_max_fit, max_fit, data, blade)
% Uses the best individuals to breed another population

num_indiv = data.no_indiv; %extract from data structure
num_genes = data.no_genes;

%calculate fitness score
score = data.a_power_in/max_fit(1)*blade.fitness(:,1) + ...
        (1-data.a_power_in)*max_fit(2)./blade.fitness(:,2);
```

```
score(i_max_fit(1)) = 1.0;
score(i_max_fit(2)) = 1.0;

%Initialise Populations
old_chord = blade.chord;
old_twist = blade.twist;
init = zeros(num_indiv,num_genes);
basis_chord = init; basis_twist = init; u_chord = init; u_twist = init;
l_chord = init; l_twist = init; int_mask = init; inv_mask = init;
rotating_index = (0:1:num_indiv-1);
init=zeros(num_indiv,1):
rotated = init; basis_index = init; u_index = init; l_index = init;
%calculate mask to for gene swapping
int_mask = rand(num_indiv,num_genes) < data.crossover;
basis_index = randperm(num_indiv);
rotated = rem(rotating_index+1,num_indiv);
u_index = basis_index(rotated+1);
l_index = u_index(rotated+1);
basis_chord = old_chord(basis_index,:);
basis_twist = old_twist(basis_index,:);

u_chord = old_chord(u_index,:); u_twist = old_twist(u_index,:);
l_chord = old_chord(l_index,:); l_twist = old_twist(l_index,:);
% Generate trial blades
trial_chord = basis_chord + 0.8*(u_chord - l_chord).*int_mask;
trial_twist = basis_twist + 0.8*(u_twist - l_twist).*int_mask;
% Constrain twist and chord of new blades
trial_chord = min(trial_chord, data.max_chord);
trial_chord = max(trial_chord, data.min_chord);
trial_twist = min(trial_twist, data.max_twist);
trial_twist = max(trial_twist, data.min_twist);
trial_fitness = survival(data,trial_chord,trial_twist);

trial_score = data.a_power_in/max_fit(1)*trial_fitness(:,1) + ...
           (1-data.a_power_in)*max_fit(2)./trial_fitness(:,2);
% Substitute trial blade for unfit or old blades
index = find(trial_score > score | blade.age == data.max_age );
if (~isempty(index))
      blade.chord(index,:) = trial_chord(index,:);
      blade.twist(index,:) = trial_twist(index,:);
      blade.fitness(index,:) = trial_fitness(index,:);
      blade.dom(index) = deal(true);
      blade.age(index) = 1;
end
```

For each iteration, the essential task is the calculation of the objective functions for C_P and T_s. This is done member by member in the function fitness:

```
function fitness= survival(data,bladechord,bladetwist)
delt = min(data.t_max/50, 0.25); % Time increment for starting calcs
for i=1:data.no_indiv;
   cp(i)=power_calc_opt(data,bladechord(i,:),bladetwist(i,:),0);
   J = inertia(bladechord(i,:),bladetwist(i,:),data);
   Ts(i)=start_calc_opt(data,bladechord(i,:),bladetwist(i,:),J, delt);
end
fitness=[cp' Ts'];
```

The fitness array comprises the transposes of the arrays for C_P and T_s. The two blade element functions for calculating power and starting time are very obvious modifications of those presented previously. J is estimated as in Chap. 6. Most of the changes were made to the starting and power programs were aimed at reducing the execution time. Using Matlab's profiler, it was found that nearly 80% of the execution time was spent in the power calculation. Because of its iterative nature, the calculations could not be vectorised. The main changes were in determining the lift and drag. The experimental data for each *Re* was fitted using least squares to generate a data file with equal increments in α, and the same minimum and maximum α. Matlab interpolation routine interp1 in LandD.m in Chap. 5 was replaced by a faster, special purpose linear interpolation in the blade element program rather than a separate function. These changes resulted in significant time savings but the execution time for the calculations in the next section was between 1 and 2 h on a moderately powerful laptop.

Further reductions in time could be achieved by parallelising the calculations as the calculations for each blade are independent of all others, but this has not been explored.

7.4 Example Blade Design: A 750 W Turbine

Figure 7.1 shows the characteristics of a typical small permanent magnet generator (PMG): the 500A unit made by Ginlong Technologies.[2] This section describes the use of the numerical optimisation routines of the previous section to design blades for that generator.

Three blades are to be used. This is the common, but not universal number, as seen from Chap. 1. The main arguments in favour of three blades are:

- The aesthetic "criteria" that three blades are more visually appealing than, say, two. However, this is more the case for large, slow turning turbines, than fast, smaller ones, and
- The gyroscopic moments during yaw (as the blades rotate) are lower in magnitude and more constant for three rather than two blades. This issue is revisited in Chaps. 8 and 9 where it s seen that gyroscopic effects are significant for turbine safety.

Against these arguments are the facts (1) that three blades are 50% more expensive than two, and (2) it was shown in Chap. 5 that the choice of N does not greatly influence the maximum C_P but changes the value of λ at which this maximum occurs. Nevertheless, convention will be followed for this example.

The rated wind speed is set at 10 m/s. This is not an obvious choice and some discussion is necessary, partly because there is a wide range of rated speeds quoted for commercially-available small wind turbines. From Fig. 1.8, it is clear that most

[2] www.ginlong.com

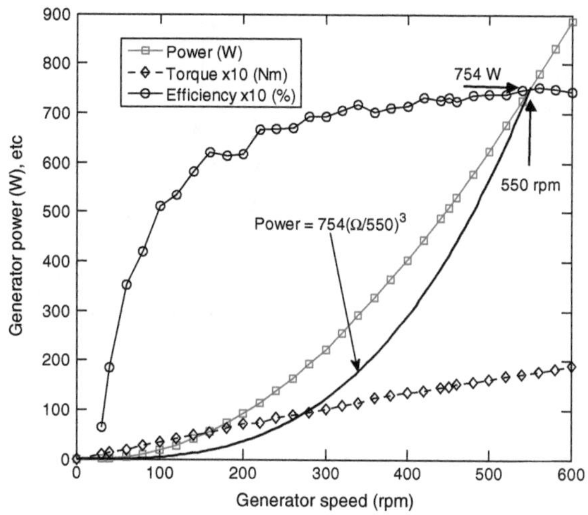

Fig. 7.1 Characteristics of the Ginlong 500 A permanent magnet generator, with rated conditions and optimum power trajectory for design example

power is produced at wind speeds above the average for the particular site. If the average is 5 m/s—a good value for small turbines—then no reduction in power due to control actions, or through braking or furling, should occur at less than about 10 m/s. For the remaining discussion of rated wind speed, it is assumed that the power curve stays flat at the desired rated power above the rated speed. Obviously, for constant rated power, lowering the rated speed will produce more power on average. However this usually requires increasing the blade radius and cost. Ω must remain constant to maintain power for a particular generator, see Fig. 7.1, so λ must increase and will soon exceed the range for optimum performance as seen in Chap. 5. Thus a choice of around 10 m/s is a compromise, but a reasonably common and often very sensible one. Note that rated aerodynamic torque is not changed by a change in rated wind speed so the starting issues, exemplified by the ratio of starting to rated torque are not strongly influenced by the choice of rated wind speed.

Table 7.1 lists the turbine parameters for the optimisation. Those shown bold are set by the choice of generator which has a maximum cogging torque of 0.5 Nm.[3] The case of no cogging torque will be studied for comparison. Furthermore, it is assumed that the hub radius is the outer radius of the generator. The blade radius, tip speed ratio, and maximum power are now discussed in terms of Fig. 7.1. The choice of maximum power as being close to the maximum available from the generator is obviously important in reducing cost, but also the higher the power, the lower the ratio of cogging to rated torque. However, the generator's maximum power should be greater than the maximum design power of the turbine, for safety and to allow some leeway for the control system. For the Ginlong generator a rated power of 754 W at

[3] Note that the author's measurements of two of these generators gave $Q_R = 0.35$ Nm, see Table 1.8.

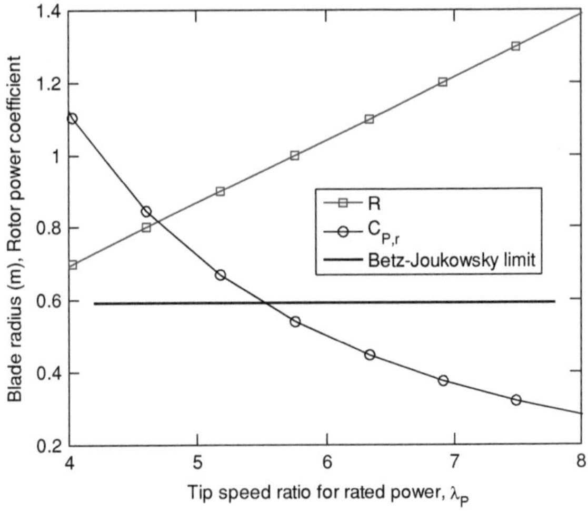

Fig. 7.2 Variation in power coefficient and blade radius with tip speed ratio for rated power

550 r.p.m. is a good choice on these grounds. Further comments on electrical system safety can be found in Chap. 11. The curve shown in Fig. 7.1 as a solid line from this point is the operating trajectory of the turbine assuming constant-λ and constant (and hopefully maximum) efficiency operation. Thus the generator will always operate below the maximum torque curve which is normally set by limits to heat generation and temperature. Note, however, that most generators can safely produce excess power for limited time. With rated Ω and power fixed at those values, the main blade parameters can be expressed as functions of R:

$$C_{P,r} = \frac{754/0.74}{\frac{1}{2} \times 1.2 \times 10^3 \times \pi R^2} = 0.541/R^2$$

and (7.3)

$$\lambda_P = \frac{550 \times 2\pi}{10 \times 60} R = 5.759R$$

Note that the subscript "r" has been added to indicate that the power coefficient refers to the power extracted by the blades not the output electrical power. The calculation of $C_{P,r}$ includes the efficiency of 74% from Fig. 7.1 to make it directly applicable to the blade element calculations. The variation of R and $C_{P,r}$ with λ_P are shown in Fig. 7.2. Again, it can be argued that the choice of minimum R is the best, but it is wise not to expect too high a C_P from a small turbine; recall the data in Table 1.1. Choosing $R = 1.06$ m gives $\lambda_P = 6.10$ and $C_{P,r} = 0.481$ from Eq. 7.3. The value of λ_P is unremarkable but that for $C_{P,r}$ is an ambitious design target. It gives an output power coefficient C_P of 0.361.

Of the values in Table 7.1 that have not yet been discussed, the choice of starting λ and blade density will be familiar from Chap. 6. Neither is critical to the

Table 7.2 Optimisation parameters

Parameter and value	Parameter and value
Population = 2000	Number of blade elements = 15
No. generations = 400	Crossover factor, CR = 0.1
Max. blade age = 20 generations	

design as the starting performance will be judged relatively. The limits on blade chord and twist are to be viewed as typical limits imposed, say, by the manufacturing process (Table 7.1). Table 7.2 lists the optimisation parameters.

A snippet of an optimisation run is shown below with the output from the data structure data suppressed

```
>> data=blade_opt_setup;
>> blade_deopt(data)
gen. Max. Cp for blade       Min. Start Time for blade
  1  0.394    972              1.026          1105
  2  0.394    972              1.026          1105
  3  0.410    127              0.936           999
  4  0.410    127              0.904          1676
  5  0.428    323              0.904          1676
  6  0.428    293              0.904          1676
  7  0.430   1583              0.904          1676
  8  0.444   1377              0.904          1676
  9  0.445   1379              0.904          1676
 10  0.445   1379              0.904          1676
 11  0.450    147              0.904          1676
 12  0.463   1680              0.904          1676
 13  0.463   1680              0.904          1676
>>
```

which shows the increase in C_p and decrease in T_s with generation number.

Figures 7.3 and 7.4 show the evolved chord and twist distributions for $Q_r = 0$ and 0.5 Nm respectively, for the values of a indicated on the figures. Figure 7.3 shows that the results for $a = 1$ with no tip losses generally follow the distributions of Eqs. 5.12 and 5.13. The DE optimisation reproduces the analytical results for this special case.

The solid and dashed lines are the optimal chord and twist from Eqs. 5.12 and 5.13 as explained in the captions. Reproducing the analytical optimisation in the limiting case gives confidence that the evolutionary strategy does find the optimal fitness front in general. There is some scatter in the numerical results, particularly near the hub for the high values of a, because the hub region contributes little to the power and, therefore exerts little evolutionary pressure on the developing design of power-producing blades. As a reduces, and starting becomes more important, the scatter reduces. Figure 7.3 for $Q_r = 0$ shows some increase in the chord near the hub, and a more noticeable increase in the twist. The tip region

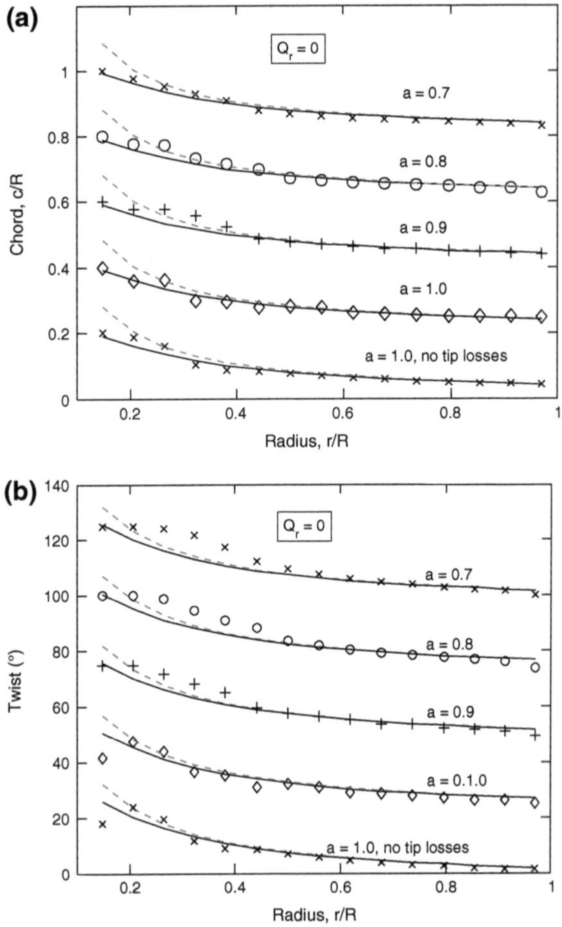

Fig. 7.3 **a** Chord distributions for average of the non-dominated blades for the values of a indicated. *Solid line* shows Eq. 5.12b and the *dashed line* shows (5.12a). Successive plots are displaced upwards by 0.2. **b** Twist distributions for non-dominated blades for the values of a indicated for $Q_r = 0$. *Solid line* shows Eq. 5.13b and the *dashed line* shows (5.13a). Successive plots are displaced upwards by 25°

hardly changes showing that starting and power production are dominated by opposite ends of the blade.

For $Q_r = 0.5$ Nm, the changes are larger and much more extensive which indicate the significance of the resistance; note that this generator has a high ratio of cogging to rated torque from Fig. 1.12.

Figure 7.5 shows the optimal fitness front for both values of resistive torque and clearly demonstrates the significance of Q_r in determining the starting time: furthermore, for $Q_r = 0.5$ Nm the BET optimum blade did not start. For $Q_r = 0$, the blade for $a = 0.8$ with $C_P = 0.484$ would be suitable for the chosen generator, provided tight control was kept over the blade shape during manufacture to ensure that the aerofoil profile was reproduced accurately. In terms of the most efficient blade, the chosen design produces 4% less power but its starting time is 33% lower. For $Q_r = 0.5$ Nm, the blade for $a = 0.9$ with $C_P = 0.487$ would be suitable.

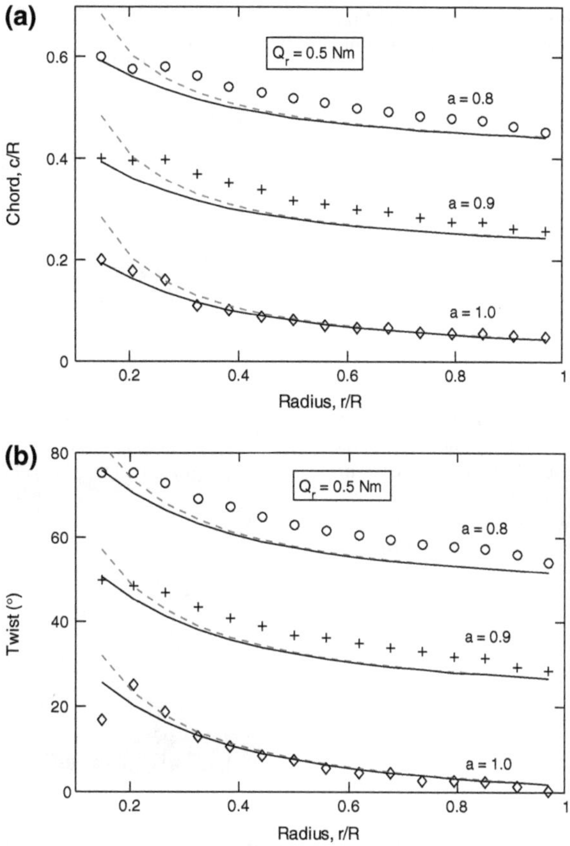

Fig. 7.4 **a** Chord distributions for non-dominated blades for the values of a indicated for $Q_r = 0.5$ Nm. *Solid line* shows Eq. 5.12 and the *dashed line* shows (5.12). Successive plots are displaced upwards by 0.2. **b** Twist distributions for non-dominated blades for the values of a indicated for $Q_r = 0.5$ Nm. *Solid line* shows Eq. 5.13 and the *dashed line* shows (5.13). Successive plots are displaced upwards by 25°

The inertia of the three blades for this design is 0.488 kgm^2, which obviously neglects the small contribution from the blade attachment. This is much greater than the generator inertia given in Table 7.1. This large difference has been mentioned several times in previous chapters and is partly the reason why the starting of a turbine with no resistive torque is independent of N, see Chap. 6.

A designer particularly keen to improve low wind performance would probably use the $a = 0.8$ blade, with 4% less power for a further 8% reduction in starting time. By slightly increasing R, and redoing the optimisation, it would be possible to further reduce the starting time at modest reduction in efficiency. Nevertheless, the best trade-off between power extraction and starting is limited to $a \geq 0.9$ approximately, for both values of Q_r.

An actual example of a dual-optimised blade is shown in the top part of Fig. 7.6. It is the 2.5 m long blade designed by the author for the two-bladed Aerogenesis 5 kW wind turbine. The bottom part of the figure shows a 61.5 m LM Glassfiber blade for large three-bladed machines. The smaller blade has significantly greater chord over the whole blade—recall from Chap. 6 that optimum

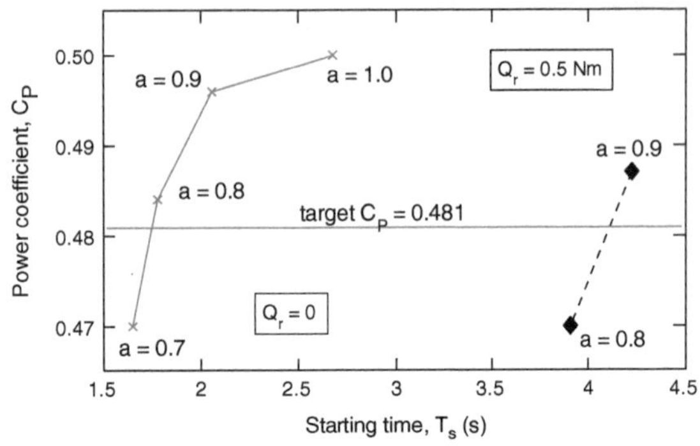

Fig. 7.5 The optimal fitness front for the blade design example

Fig. 7.6 The Aerogenesis 2.5 m long blade compared to the LM Glassfiber 61.5 m blade from www.lmwindpower.com (accessed 4 Aug 2010)

power requires *Nc* to be roughly constant—but the biggest differences in shape occur at the tip and at the hub. Near the hub, the wider chord improves the starting behaviour with two blades and relatively high resistive torque. Sharp tips on the large blades reduce the noise associated with the formation of the tip vortex: Oerlemans et al. [2] found that modern tips like that shown in Fig. 7.6 effectively eliminate tip noise.

Calculations by Clifton-Smith [3] suggest that tip noise is not a major noise source for small blades provided the tip is rounded. However, as power level increases, more attention should be paid to the tip design to reduce noise. Clifton-Smith [3] included noise as one objective function for the design of 3 m diameter blades. He found that the BET optimum blade was the noiseist, but that a significant reduction in both starting time and sound power level were achievable with only a small drop in C_P.

Determining the basic blade shape from aerodynamic optimisation is the essential first step in the design process but there is still considerable work to do to produce an actual blade that performs as designed and is structurally sound. At this point the blade connection should be considered, bearing in mind that the thick sections seen near the hub on the large blade in Fig. 7.6 are to be avoided for small blades. Chapter 4 explains the reasons for this statement which are

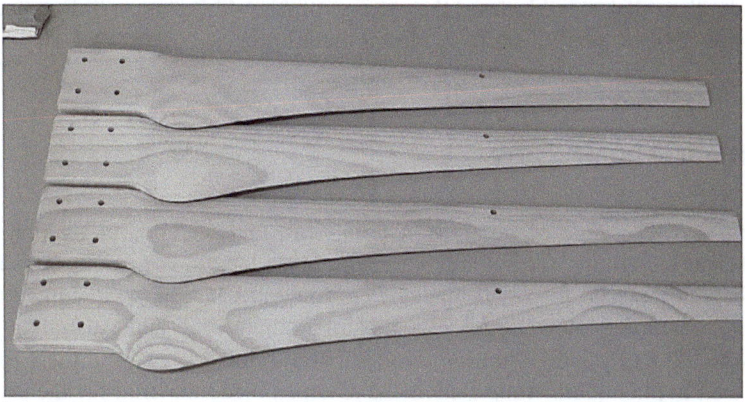

Fig. 7.7 Machined 0.87 m long timber blades for the turbine in Fig. 6.1. These blades have holes at the 2/3 radius point for static and fatigue testing

Fig. 7.8 Finite element model of the Aerogenesis 2.5 m long blade for Load Case H of the IEC simple load model. The strain is colour-coded with the maximum occurring in the constant chord region at the hub. Image from Phil Clausen

reiterated at the start of this chapter. Most small blades are held in a rectangular attachment section similar to that shown in Fig. 6.1. The back plate of the blade holder is attached to the generator shaft and then four bolts sandwich each blade between the front and back plates. These bolt holes can be seen in Fig. 7.7. For commercial turbines it is advisable to make it difficult or impossible to mount the blades backwards for protection against inexpert installation. The attachment section often lies in the plane of rotation, although not always. The two blades on the Aerogenesis 5 kW turbine shown in Fig. 7.6 are mounted at 20° to the plane of rotation. Extensive finite element analysis (FEA) as demonstrated in Fig. 7.8 showed that this arrangement minimises the stresses in this region. The other noteworthy feature of the small blade in Fig. 7.6 is that its leading edge is straight. Recall that blade element analysis does not fix the element position along the chord line so a wide range of tapers is possible for the leading and trailing edges for the same net taper of the blade. The straight leading edge was selected on the grounds that this would ease the fitting of the fibreglass reinforcement if composite blades were made by a closed mould process.

7.5 Blade Manufacture

There is a wide range of possible materials for, and methods of making small blades. The suitability of each may vary with blade length. For example, timber is an excellent material for small blades. Peterson and Clausen [4] documented the material properties and fatigue behaviour of Radiata Pine and Australian Hoop Pine which is extensively used for ultralight aircraft propellers. Both these timbers grow well in plantation. Hoop Pine is the assumed blade material for the example blade design in the previous section and the application of the IEC SLM in Chap. 9. Sinha et al. [5] provide similar data for Nepali timbers as well as information on weathering of a number of surface finishes. Other timbers, including Sitka Spruce, also widely used for propellers, and Douglas Fir, are suitable for wind turbine blades, provided they are grown sustainably. However, the cost of laminating it is high, and so the only practical way of using timber is to carve or machine blades from solid blanks. As length increases, it becomes more difficult to obtain blanks that are knot- and defect-free. Computer numerically controlled (CNC) milling of timber, which was used for the 0.87 m long blades in Fig. 7.7, is straightforward but is probably too expensive for volume production. The most promising technique would appear to be the use of a copying router using a master blade possibly cut from metal on a regular CNC machine. Development of this method is underway at the University of Calgary and progress will be reported in the online materials http://extras.springer.com. Whatever method is used, considerable care is required to reproduce the design aerofoil profile. Figure 7.9 shows measurements of the surface of one of the blades in Fig. 7.7 near its tip, in the region where significant differences in blade shape could have large impact on the power output. Unpublished calculations by Barbara van Bossuyt using aerofoil computational programs XFOIL and RFOIL did not indicate a significant power loss due to the change in shape (Fig. 7.9).

For longer blades some form of composite manufacture is preferable for which there are many material and manufacturing issues shared with large blades, e.g. Brøndsted et al. [6] and Dutton et al. [7]. For example, blades of all sizes

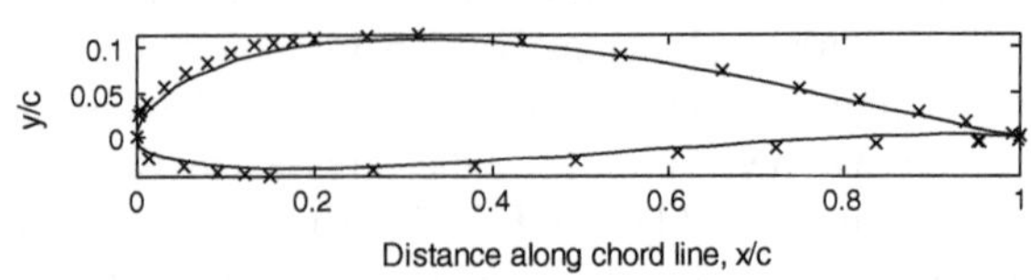

Fig. 7.9 Measured blade surface at $r/R = 0.995$ on one blade from Fig. 7.7. The *crosses* show the measured surface. The design aerofoil profile (SD7062) is the *solid line*

Fig. 7.10 Machining the mould for the lower surface of the 2.5 m long blade in Fig. 7.6

require adequate stiffness to avoid excessive bending under load, and a low inertia. For large blades the latter relates to manufacturing and transport costs, whereas for small blades low inertia is more important for starting. There is an enormous number of manufacturing techniques available so it possible only to give a broad and general outline. As for large blades, it is desirable to concentrate the structural strength in a thin laminate on the blade surface. The laminate is made by embedding reinforcing material such as fibreglass or carbon fibre within a resin. Because of the high level of centrifugal stress, see Sect. 1.9, considerable reinforcing in the radial direction is required, often in the form of "unidirectional" E-glass. In addition, "triaxial" reinforcement in the so-called 45°/90°/45° directions, where 90° indicates radial reinforcement, is often used for torsional stiffness. The reinforcement layout—the number, type of material, and its location and extent—requires detailed FEA as demonstrated in Fig. 7.8. In turn, the FEA must be carefully checked as discussed in the next section.

Moulds are needed for composite blade manufacture. For large blades, these are often made by machining thin templates which are then spaced along the span with the gaps filled with resin and reinforcement. For small blades, dimensional accuracy is critical, so machined moulds are often necessary. The machining of one of the moulds for the 2.5 m long blades in Fig. 7.6 is shown in Fig. 7.10.

Separate moulds were made for the lower and upper surface and each blade half made by vacuum infusion, Fig. 7.11. After the moulds are coated with a release agent, the fibreglass is laid out by hand in the mould, covered by a "release ply", a "resin runner", and vacuum bag. The fibreglass is not visible in Fig. 7.11 as it

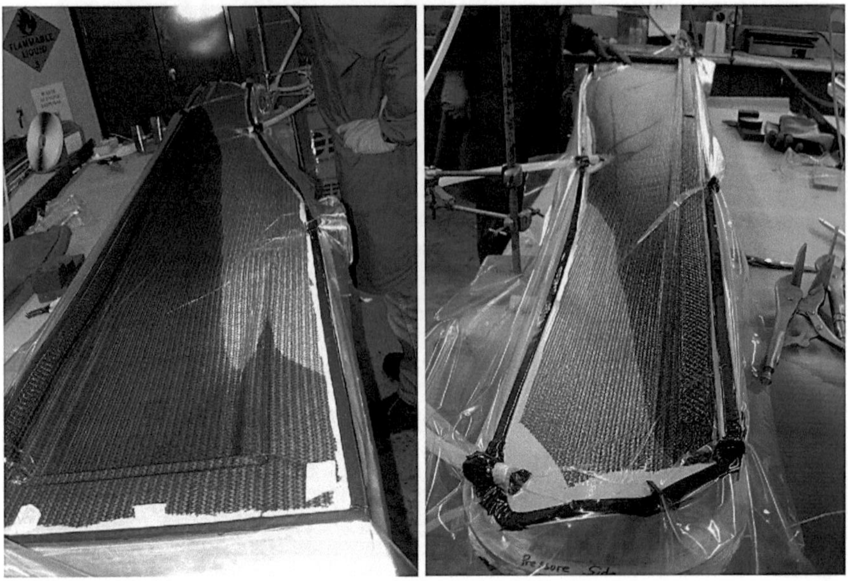

Fig. 7.11 Vacuum infusion of the lower half of a 2.5 m long blade. *Left side* view from hub end, *right side* from the tip

and the release ply are covered by the (green) resin flow promoter which is cheap and readily-obtainable shadecloth. The release ply is visible at the bottom of the right hand side photograph in Fig. 7.11. A combination of the promoter and cheap "geocloth" forms the resin runner along the leading edge. This promotes the rapid flow of resin along the blade which is essential for the high aspect ratio shape. The vacuum bag is sealed by the black sealer running around the flat landing of the mould and the resin inlet port is just out of the left side picture on the extreme left. The two white exit tubes are clearly seen at the top of the photograph on the left and towards the bottom on the right side one. They are connected to a vacuum pump which sucks the resin through the mould. The dark (green) area is the region wetted by the resin. When the whole blade is wetted out—hopefully just before all the resin has entered the mould—the inlet and outlet ports are pinched off and the resin allowed to cure. It may be necessary to heat the blades halves to promote curing.

When cured, the release ply and resin runner are peeled from the laminate. Then the two halves must be trimmed and joined. This can be a time-consuming process and one that must be done carefully as the blade halves will likely join at the leading and trailing edges, which are aerodynamically important. The trailing edge is relatively easy because the two halves present a good area for gluing, but the laminates, which may be only 1–2 mm in thickness, meet at nearly 90° at the leading edge. It is important to have some form of "glue dam" to ensure good bonding. Measurements of the surface of a blade made by vacuum infusion are shown in Fig. 7.12. It is clear that the blade is too thin in comparison to the design

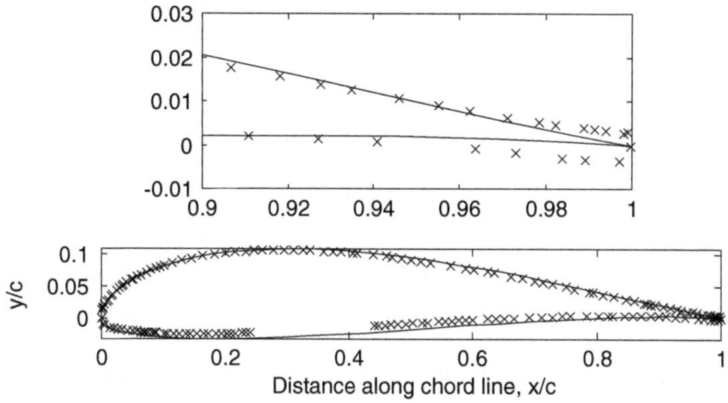

Fig. 7.12 Measured blade section of the Aerogenesis 2.5 m long blade from the power producing section. *Bottom graph* shows whole section. The *top graph* highlights the trailing edge (data from Rachel Wong). The gap in the lower surface data is due to mounting of blade section for measurement

aerofoil, probably as a result of the way the two halves were joined. The upper surface (with most curvature) was placed in its mould and the lower surface was vacuumed onto it after applying the glue. It is thought that the vacuum deflected the lower surface towards the upper one. The upper part of the figure shows an expanded view of the trailing edge, which clearly demonstrates the technique of manufacture. Trailing edge thickness can be a problem for small blades because of the conflicting requirements of finite thickness for safety and integrity and minimal thickness to reproduce the design shape and suppress trailing edge noise. This arises from vortex shedding behind a trailing edge of finite thickness and is usually correlated by the ratio of trailing edge thickness to the chord, it is difficult to keep this parameter small for small blades. Nevertheless, noise calculations by Clifton-Smith [3] suggest that trailing edge thicknesses of 1–3 mm are tolerable for most small blades. No computational studies have yet been done to study the effect of the change in shape on turbine power extraction.

Significant parts of the upper (downwind) surface of any blade will be in compression for most of the operating life of the turbine, so it is possible that a blade can buckle. Partly to prevent this, most large blades have a "shear web" or "spar", as discussed, for example by Burton et al. [6], Brønsted et al. [7], and Dutton et al. [8]. This is a short (in the chord direction) box keeping the blade halves apart in the region of maximum thickness. Not all small blades have a spar; the blades of the Skystream shown in Fig. 1.2 are hollow. If used, the spar needs to extend to the leading edge to be a glue dam or some other arrangement is necessary.

It should be clear from this description that vacuum infusion requires care and considerable labour and so may not be suitable for volume production by poorly-trained workers. On the other hand, its capital cost is small, at least once the moulds have been provided, so it a good method to make a small number of blades. Higher volume production often uses a closed-mould process, such as some form of resin-transfer moulding (RTM). RTM (usually) requires a core which is (usually) made separately and then placed in the mould along with the fibreglass. The mould is then closed and resin injected under pressure. If the mould is sealed correctly there is little trimming required. It was to make RTM easier that the leading edge of the blade in Fig. 7.6 was made straight.

There are many variations possible on the themes of vacuum infusion and resin transfer moulding, such as light-RTM, vacuum assisted RTM etc. and it is possible to make laminates using pre-impregnated or "prepreg" reinforcement. Furthermore, it is possible to form the core at the same time as the laminate surface. There is also a considerable choice in the resins that can be used. The most common for large blades are vinylester, polyester, and epoxy resins supplied by companies like Huntsman[4] and Hexion.[5] Makers of small blades should seriously consider using resins developed for large blades as long as they comply with guidelines such as GL [9]. If used properly any large-blade resin will have material properties that are appropriate for small blades. Epoxy resins tend to more expensive but the difference in terms of the total material costs for a blade is likely to be small. They also tend to have the longest shelf life and superior fatigue properties. Polyester resins are usually easier to handle, but have a higher shrinkage and curing temperature. The blades in Fig. 7.6 were made using vinylester. However the shrinkage caused "print through" as the blade surface conformed to the shape of the reinforcement immediately below it rather than the mould surface, so an epoxy resin was substituted for later blades. The key resin properties are the viscosity, which determines how well it flows and wets out a mould, pot life, which limits the time during which a blade can be made, and the glass transition temperature, which determines the curing temperature. GL [9] requires a glass transition temperature of at least 65°C, and higher than any operational temperature. The GL guidelines also specify that the material safety factors for composite small blades must take account of environmental degradation and operating temperatures. Safety factors are further described and used in Chap. 9.

It is one of life's ironies that the resins used for wind turbine blades are all derivatives of petroleum. This has lead to considerable interest in sustainable alternatives, of which timber is obviously one. Other sustainable materials such as bamboo are being investigated as alternatives to current reinforcements, [10]. Developments could also include more widespread use of timber spars for small blades, perhaps using Paulownia, a low density Asian timber with a straight grain that grows rapidly in plantation.[6]

A wide range of gelcoats can be used when moulding blades to give a good surface finish.[7] Some resins are more suited than others to gelcoats: epoxy resins,

---

4 http://www.huntsman.com/advanced_materials/ and click on "Wind Energy" (accessed 2 Sept 2010).

5 http://www.hexionchem.com/Industry/wind_energy.aspx?id=8174 (accessed 2 Sept 2010).

6 http://www.worldpaulownia.com/ (accessed 2 Sept 2010).

for example, normally require a primer to bond to a gelcoat. Epoxy resin is also particularly sensitive to ultraviolet degradation and so must be covered if a 20 year blade life is required. Again there are several choices, including tough two-part epoxy paints. Leading-edge tape can provide additional protection in the region where insect and dust impact is greatest[8] or materials like Bladeskyn[9] can be used to coat the whole blade surface.

7.6 Blade Testing

The blades are the most critical component of a wind turbine in terms of safety, so it is important to test blades in as many ways as possible. The main tests are:

- "coupon" tests of the blade material to determine material properties like Young's modulus and the yield stress, verify FE models, and justify safety factors over operational temperature ranges,
- static blade tests, and
- fatigue tests

Coupon tests are specifically required by GL [9] to establish the temperature-dependence of the blade material properties. However, only the second of the three listed tests is required by IEC 61400-2. Figure 7.13 shows a static test of a Hoop Pine blade of the shape shown in Fig. 7.7. Clearly Hoop Pine is a very suitable blade material. In a static test, it is important that the blade be mounted as it is on an actual turbine. These simple tests measure deflection against load, and provide an important check of the blade structural modeling such as the FEA in Fig. 7.8. This ensures, for example, that a blade bending during high winds does not hit the tower. Load Case H in the IEC SLM covers this situation as described in Chap. 9. Generally small turbines have a larger distance between the tower axis and the blade attachment point than do large turbines, when expressed as a fraction of R, so such contact is less likely. Nevertheless it is still necessary to demonstrate clearance.

Extra information is obtained from static torsional tests and determination of the blade's natural frequency. With this information it has been demonstrated that FE models can accurately predict the important operational loads on small wind turbines, e.g. [11–13].

Using the measurements of Bechly and Clausen [11, 12] from an operating, strain-gauged blade, Epaarachchi and Clausen [14] developed a fatigue load procedure to simulate a full life of an operational blade, typically 20 years. Figure 7.14 shows the 5 kW, 2.5 m long blade from Fig. 7.6 in a fatigue test rig

---

[7] http://www.scottbader.com/composites-products-gelcoats.aspx (accessed 3 Sept 2010).

[8] http://www.aircraftspruce.com/catalog/cspages/leadingedgetape.php (accessed 3 Sept 2010).

[9] http://www.bladedynamics.com (accessed 4 Sept 2010).

Fig. 7.13 Static test of a 0.87 m long timber blade at the University of Newcastle (photograph by Paul Peterson)

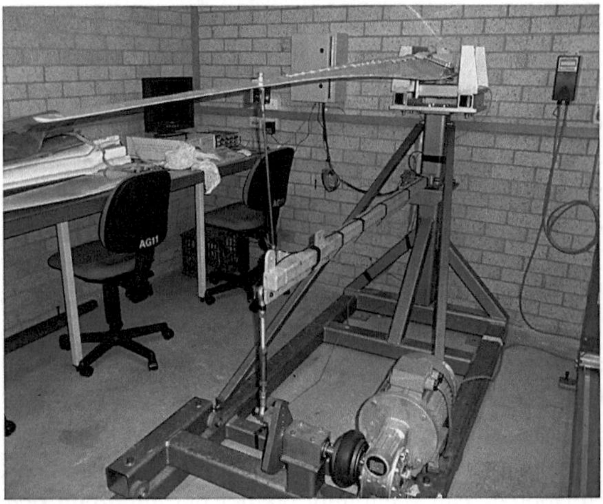

Fig. 7.14 The 2.5 m long blade from Fig. 7.6 in the fatigue test rig at the University of Newcastle

ready for testing. The electric motor at the bottom right is controlled by the variable frequency drive at the top right. The blade is shaken by the hinged horizontal arm attached via a vertical rod at the 2/3rd radius point, chosen as being close to the centre of pressure for thrust loads that are proportional to radius, such as those shown in Fig. 5.6. Amplitude variation requires a manual adjustment to the vertical, eccentrically-mounted arm in the bottom centre of the photograph. Note that the blade is mounted at an angle to the horizontal: this angle is also its pitch angle on the actual turbine. A number of strain gauges are mounted along the blade. Their output is collected by the data logger in the cabinet mounted on the wall behind the blade. Monitoring these gauges gives a good record of the development of fatigue damage because any change in the blade properties due to fatigue will alter the strain.

Previous versions of the blade were mounted in the plane of rotation (or horizontally in the rig shown) in much the same way as the smaller blades in Fig. 7.7 are mounted in the turbine shown in Fig. 6.1. However, the stress concentrations in the transition region between the attachment and aerodynamic portion of the blade induced delamination and fibre pullout on the tension (lower) surface after the equivalent of only 7 years. Straightening the blade achieved superior strength with fewer fibreglass layers at the cost of complicating the blade attachment. It is not clear from the photograph that the blade holder is held by two load cells to monitor the force and moment on the blade. This is not strictly necessary but it provides information about the relationship between constant amplitude testing, which is done for convenience, and constant force testing, which is likely to be more realistic.

The fatigue test program consists of a number of cycles at various frequencies and stroke. Some of the "bins" of cycles are amalgamated if their parameters are similar, and the tests are done partly in multi-year groups of single cycles. Both these simplifications reduce the number of manual adjustments to the stroke. After the photograph in Fig. 7.14 was taken, the blade survived undamaged an accelerated test program simulating 20 years of operation.

Fatigue behaviour of composite wind turbine blades is a complex subject, e.g. [15], and is probably deserving of a book on its own. Some basic and very useful information on the fatigue behaviour of composites and timber is given in Annex E of IEC 61400-2 which should be consulted by all blade designers. It is likely that fatigue testing will become compulsory under future revisions of the IEC standard and this will be accompanied by an increased knowledge of the damage mechanisms and techniques to design against these. In the meantime it is important to be careful and conservative in assessing small blade fatigue.

7.7 Forming the Rotor

Once blades have been made and tested for suitability and integrity, and the blade holder designed and manufactured, the blades can be attached to form a rotor. This process is usually straightforward, especially for small turbines without pitch control, but there are a few issues that must be considered. The main one is to decide on the maximum allowable imbalance of the rotor which is likely to be the largest imbalance in the whole turbine and a potential source of vibration and fatigue. IEC 61400-2 assumes a default eccentricity in the rotor centre of mass, e, of $0.005R$ for the SLM as explained in Chap. 9. For a two-bladed rotor, e is given by

$$e = \frac{|d_1 - d_2|}{M} \tag{7.4}$$

where M is the sum of the blade masses and d is the product of the blade mass and centre of mass. Equation 7.4 holds only if the blade masses lie in the same plane and are exactly 180° apart. For $N = 3$, the formula is

$$e^2 = \frac{1}{2M^2}\left[(d_1 - d_2)^2 + (d_1 - d_3)^2 + (d_2 - d_3)^2\right] \tag{7.5}$$

provided the blades are planar and exactly 120° apart.

There is very little publicly-available information on actual and allowable values of e. Ramlau and Niebsch [16] describe a large rotor with $e \approx 0.003R$ suggesting the value of $0.005R$ is a reasonable upper limit. The author's experience supports this.

If the blades are made in a batch larger than N the question then arises of how to optimally form the rotors in such a way as to minimise the sum of e^2 over all rotors. This problem is addressed by Hitz and Wood [17] to which the interested reader is referred for more details. They found that when the standard deviation in mass and centre of mass was 1% or less, optimal matching is nearly equivalent to ordering the blades in terms of d. One fascinating aspect of the problem is that (7.4) and (7.5) are unchanged if the blade indices are permuted. However, this is not the case for $N \geq 4$ and the optimisation of blade matching for large N is far from simple, if not a major problem for mainstream turbines.

7.7.1 Exercises

1. The discussion at the start of Sect. 7.4 of the advantages and disadvantages of three rather than two blades was brief and ignored several important considerations. What are the missing considerations?
2. Interpret the optimal power trajectory in Fig. 7.1 in terms of the performance curve such as Fig. 5.2a by ignoring Reynolds number effects.
3. The generator used for the blade design example apparently shows a significant decrease in efficiency as power decreases. How would this alter the trajectory in Fig. 7.1?
4. The blade section manufacturing errors in Figs. 7.9 and 7.12 may have an effect on power extraction but should have little effect on starting performance. Is this statement correct?
5. Use the internet and other sources to find out what information you can on locally-available timbers that could be used for small blades. What information is required and can you find sufficient for each candidate species?
6. Derive Eqs. 7.4 and 7.5.
7. For an N-bladed rotor in which the blades are co-planar and equi-spaced, show the eccentricity of the centre of mass e is given by

$$e^2 = \sum_1^N \sum_1^N a_{lm} d_l d_m \Big/ \sum_1^N m_i \quad \text{where } a_{lm} = \cos(2\pi(l - m)/N)$$

and m_i is the mass of blade i.

References

1. Price KV, Storn RM, Lampinen JA (2005) Differential evolution: a practical approach to global optimization. Springer, Berlin
2. Oerlemans S, Sijtsma P, Mendez Lopez B (2007) Location and quantification of noise sources on a wind turbine. J Sound Vib 299:869–883
3. Clifton-Smith M (2010) Aerodynamic noise reduction for small wind turbine rotors. Wind Eng 34:403–420
4. Peterson P, Clausen PD (2004) Timber for high efficiency small wind turbine blades. Wind Eng 28:87–96
5. Sinha R, Acharya P, Freere P, Sharma R, Ghimire P, Mishnaevsky L (2010) Selection of Nepalese timber for small wind turbine blade construction. Wind Eng 34:263–276
6. Brøndsted P, Lilholt H, Lystrup A (2005) Composite materials for wind power turbine blades. Ann Rev Mater Res 35:505–538
7. Dutton AG, Bonnet PA, Hogg P, Lleong YL (2010) Novel materials and modelling for large wind turbine blades. J Power Energy 224:203–210
8. Burton T, Sharpe D, Jenkins N, Bossanyi E (2001) Wind energy handbook. Wiley, New York
9. GL (2010) Guidelines for the certification of wind turbines, Germanischer Lloyd. https://www.gl-group.com/wind_guidelines/wind_guidelines.php?lang=en. (accessed 10 Oct 2010)
10. Holmes JW, Brønsted P, Sørensen BF, Jiang Z, Sun Z, Chien X (2009) Development of bamboo-based composite as a sustainable green material for wind turbine blades. Wind Eng 33:197–210
11. Bechly ME, Clausen PD (2001) The dynamic performance of a composite blade from a 5 kW wind turbine. Part I: measured blade response. Wind Eng 25:133–148
12. Bechly ME, Clausen PD (2002) The dynamic performance of a composite blade from a 5 kW wind turbine. Part II: predicted blade response. Wind Eng 26:273–286
13. Wilson SVR, Clausen PD (2007) Aspects of the dynamic response of a small wind turbine blade in highly turbulent flow: part 1 measured blade response. Wind Eng 31:1–16
14. Epaarachchi JA, Clausen PD (2006) The development of a fatigue loading spectrum for small wind turbine blades. J Wind Eng Ind Aerodyn 94:207–223
15. Mandell JF, Samborsky DD, Wang L, Wahl NK (2003) New fatigue data for wind turbine blade materials. J Solar Energy Eng 125:506–513
16. Ramlau R, Niebsch J (2009) Imbalance estimation without test masses for wind turbines. J Solar Energy Eng 131:1–7
17. Hitz KL, Wood DH (2010) On blade matching following batch production. Wind Eng 34:325–333

Chapter 8
The Unsteady Aerodynamics of Turbine Yaw and Over-Speed Protection

8.1 Introduction

The analyses of power extraction and starting in the last six chapters assumed effectively steady flow whereas all turbines operate in unsteady flow for most of the time. The effects on turbine operation are significant and complex. They include "dynamic inflow", the general name for the unsteady behaviour of the axial and rotational induction factors, which in turn are influenced by the vortex structure of the time-varying wake, as well as the unsteady lift and drag on the blades, e.g. Leishman [1, 2]. While important, these aspects are not critical for an introductory treatment, if for no other reason than that a determination of the turbine power curve will usually reproduce the design steady performance curve if the turbine is well built and the testing is done to the appropriate IEC standard.

Unsteadiness, however, is unavoidable in examining yaw behaviour, the turbine's motion about the vertical "yaw" or tower axis, in response to changes in wind direction. This behaviour is important because a yaw "error", θ, between the turbine axis and the wind direction, reduces turbine power by the ratio $\cos^2\theta$ to a first approximation, see Sect. 1.2. Thus a 20° yaw error reduces the power by a significant 12%. In addition, one of the IEC SLM Load Cases studied in the next chapter is associated with yaw error. Most small wind turbines have a tail fin, as can be seen in Figs. 1.2 and 6.1. Their yaw response depends on tail fin area, moment of inertia, and distance from the yaw axis. The shape of the tail fin is not usually critical.

Yaw behaviour is also connected with a number of important safety issues. For example, and somewhat surprisingly at first sight, yawing can lead to significant cyclic (and hence unsteady) gyroscopic loads. It is shown in Chap. 9 that these loads are often the largest loads on the blade roots and rotor shaft of a small turbine. To a first approximation, they are of the form

$$M = k_N J \Omega \omega \tag{8.1}$$

where M is the load (moment), k_N is a numerical factor depending on the number of blades, J is the moment of inertia of the blade (see Chap. 6), and ω is the yaw rate.

D. Wood, *Small Wind Turbines*, Green Energy and Technology,
DOI: 10.1007/978-1-84996-175-2_8, © Springer-Verlag London Limited 2011

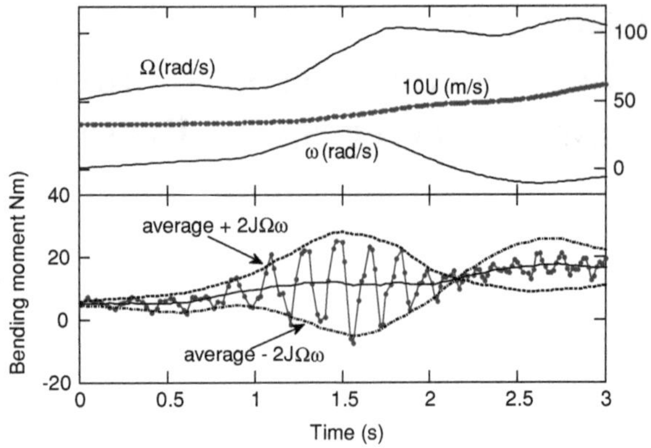

Fig. 8.1 Gyroscopic loads on blade root of the turbine in Fig. 6.1. The *top part* shows wind and rotor speed and yaw rate. *Bottom part* shows measured blade root bending moment in Nm (*solid line with symbols*), the short-term averaged bending moment in Nm (*solid line*) as explained in the text, and the envelopes of the deviation from the average as given by Eq. 8.1. The origin for time is arbitrary. Data from Sturt Wilson

Equation 8.1 gives the maximum magnitude of the cyclic gyroscopic component. Figure 8.1 demonstrates these loads for the turbine depicted in Fig. 6.1. Special blades were made, one with strain gauges embedded on the pressure (upwind) surface and the electrical connections placed inside the blade to minimise the aerodynamic interference. To maximise the strains, the blades were made deliberately weak. The blade root bending moment plotted in the bottom part of Fig. 8.1 was calculated from the measured strains. Also measured were the wind speed, yaw rate, and rotor rpm. The bending moment is caused by the aerodynamic pressure on the blade surface giving the "short-term averaged" moment and the cyclic gyroscopic load whose period corresponds to Ω. For this moment, $k_N = 2$ in Eq. 8.1, and Fig. 8.1 shows that this envelope is a good fit to the data. It is clear from Fig. 8.1 that the (unsteady) cyclic load is "activated" by the high yaw rate centred at time = 1.5 s.

To avoid large gyroscopic loads the tail fin must be designed to prevent the turbine following "high frequency" wind direction changes, but it must follow the "low frequency" wind direction changes to maximise output power. Finally, one of the major ways of protecting small turbines against over-speeding, either in high winds or when the electrical load has been lost, is furling. This involves collapsing the tail fin at hopefully, a predictable and repeatable wind speed and loading. For micro turbines, an alternative is to pitch the turbine out of the wind, a technique that also must take account of the gyroscopic moments. Larger turbines may use a brake which is discussed along with electronic speed control in Chap. 11.

8.2 Fundamentals of Tail Fin Aerodynamics

Large wind turbines have a wind vane on or near the nacelle, and use a motor to drive the turbine about the yaw axis to keep it pointed into the measured wind direction—this type of yaw system is called "active yaw". Additionally, this yaw drive can be used to turn the turbine away from the wind to limit the aerodynamic loads on the rotor during extreme events such as storms, cyclones, typhoons, or hurricanes.

On the other hand, most small wind turbines are "free yaw" systems, in which the rotor is located upwind of the generator and tower, and a tail fin keeps the turbine pointing into the wind. This simpler yaw system reduces the cost of small turbines, and is easier for the owner to maintain. An alternative to a tail fin is to have the blades downwind of the tower which can cause "tower shadow" effects and reduce the fatigue life of the blades. Despite their mechanical simplicity, tail fin aerodynamics can be complex since the yaw rate is now unregulated and at the mercy of the wind. Free yaw behaviour is largely a function of the tail fin design especially during starting when the blades provide little yaw stability.

An examination of all the small wind turbines produced over the years would show a wonderful array of tail fin designs, but most are simply variations of a flat plate acting as a wing. Swept wing designs are common and their simplest form, the delta wing, has been the subject of considerable study [3–5]. Aside from being aesthetically pleasing, easily manufactured, and strong, delta wings have favourable and well-known aerodynamic characteristics. For this reason, the theory of tail fins is developed in terms of delta wing aerodynamics. The most notable feature of delta wings is their high stall angle, which in the context of a tail fin suggests high restoring moments on the turbine up to yaw angles of around 40°. Figure 8.2 defines the basic geometry of a delta wing tail fin: The chord is c, the span is b, and the aspect ratio, AR, is $2b/c$. Theoretically, the centre of pressure on a delta wing is $2c/3$ from the apex, so the moment distance r, should be measured from this point to the yaw axis. In practice, $2c/3$ is usually small compared to the tail boom length and can often be neglected. Figure 8.3 shows the lift and drag characteristics for the delta wing closest in geometry to the tail of the 500 W turbine in Fig. 6.1; note that a delta wing is three dimensional so the lift and drag are in Newtons. This data was compiled from the experimental studies cited in Table 8.1.

There are at least two reasons why high angle data is needed even though a tail fin should operate near $\alpha = 0°$. First, an actual operation sequence of the 500 W turbine from Chap. 6 will be presented later to show that high yaw angles occur in practice, usually at low wind speeds before and during starting. Second, it is possible to have a failure of the yaw bearing mechanism which may result in excessive wind loads on a stationary turbine at high incidence. This possibility is discussed further in Chap. 9.

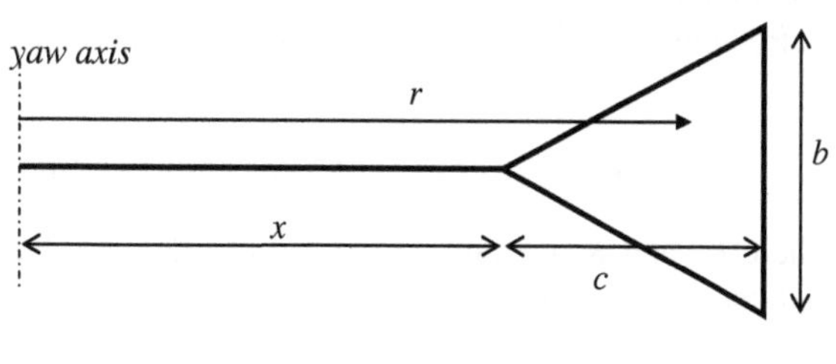

Fig. 8.2 Basic geometry of a delta wing tail fin. r is the distance from the yaw axis to the centre of pressure

Table 8.1 Symbols used in Fig. 8.3

Symbol	Source	Comments
◇	Equation 8.2	AR = 1.73
×	Kegelman and Roos [26]	AR = 1.46
*	Torres and Mueller [18]	Flat plate AR = 1.75
□	Traub et al. [27]	AR = 1.46
+	Wentz and Kohlman [28]	AR = 1.86
Δ	Matsumiya et al. [29]	AR = 2.0
O	Koenig [30]	AR = 1.0
—	Present estimate	

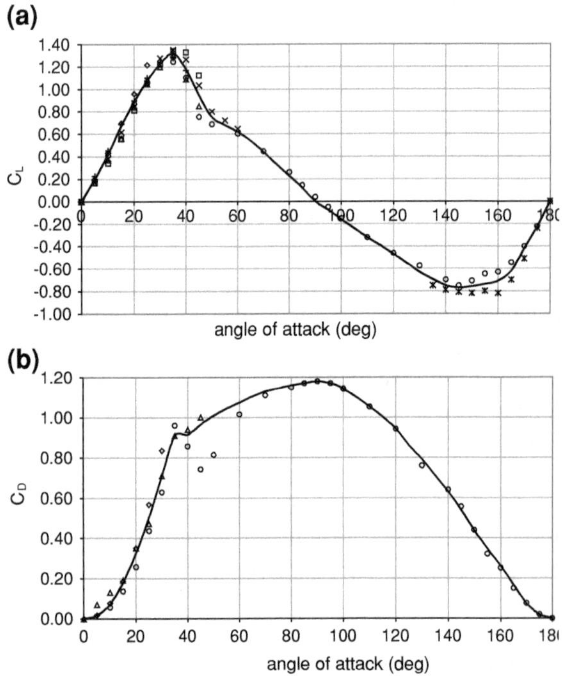

Fig. 8.3 **a** Composite approximation of lift coefficient for delta fin of AR = 1.73 from Wright [21]. **b** Composite approximation of drag coefficient for delta fin of AR = 1.73 from Wright [21]

It is generally held that the aerodynamics of delta wings are much less *Re*-dependent than that of aerofoils, mainly because the flow is dominated (for all *Re*) by the separation which is fixed at the leading edge. Otherwise, there are interesting similarities to aerofoil lift and drag in that lift is nearly linear in α at small angles, and the drag coefficient is maximised at $\alpha = 90°$.

Polhamus [6, 7] derived the following equations for the lift and drag of delta wings:

$$C_L = K_p \sin\alpha \cos^2\alpha + K_v \cos\alpha \sin^2\alpha \quad \text{and} \quad C_D = C_L \tan\alpha \tag{8.2}$$

where K_p and K_v depend on the aspect ratio, e.g. Bertin and Cummings [8]. Note that the relation between lift and drag is the same as for the high-α flat plate equations shown in Figs. 4.5 and 4.6 and implies that the pressure forces dominate over viscous forces. Equation 8.2 appears to be accurate for delta, and other shaped, wings for angles up to about 20–30°. At small α, the lift is linear in α with slope K_p.

From the quasi-steady analysis of starting performance in Chap. 6, it may be thought that predicting tail fin yaw behaviour simply involves using the lift and drag coefficients for a particular yaw angle to determine the normal force on the tail fin at a given wind speed, and then calculating the product of this force and the distance r in Fig. 8.3 to get the yaw moment acting on the turbine. However, this approach ignores the location of the tail fin in the rotor wake, and the change in lift and drag due to the changing wind direction and speed.

Consider first the wake in the vicinity of the tail fin. This is an topic of ongoing research, and there is no consensus yet on what flow speed or direction to assume when analysing tail fin aerodynamics. One assumption is that the tail fin is located in the "near wake" region of the rotor wake, so that, U_{wake}, the wind speed experienced by the tail is

$$U_{\text{wake}} = U_0(1 - a) \tag{8.3}$$

For optimum turbine operation, the axial induction factor, $a = 0.33$ and $U_{\text{wake}} = 0.67U_0$, where U_0 is the undisturbed wind speed. It is not clear whether the steady induction factor can always be used for unsteady flow.

8.3 Unsteady Aerodynamics of Tail Fins

There are at least two different ways to analyse the unsteady yaw dynamics of a tail fin, both giving the same general result. For the present, the stabilising effects of the blades are ignored. The first is to assume that static lift and drag are applicable in unsteady conditions. The angle of attack on the tail fin is found by a vector addition of the actual wind direction and the angular velocity of the tail fin about the yaw axis as shown schematically in Fig. 8.4. θ is the angle between the wind direction and the tail fin, U is the wind speed, and the subscript "a" refers to

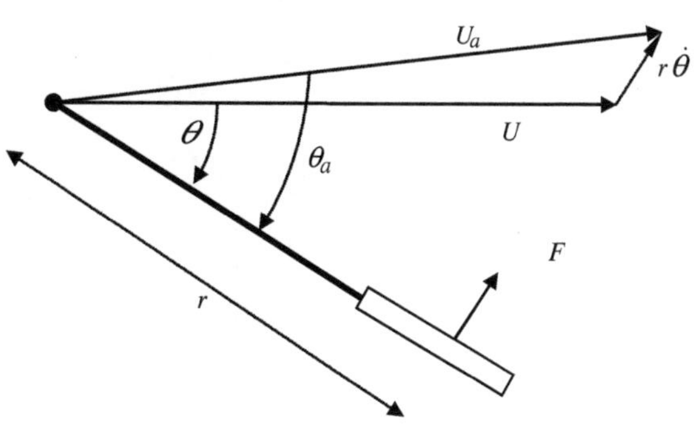

Fig. 8.4 Schematic of tail fin motion about the yaw axis

the apparent wind speed and angle of attack on the tail fin moving with angular velocity $\dot{\theta}$ under the action of aerodynamic force F. Note that $\dot{\theta}$ is not necessarily the yaw rate as the co-ordinate system used in Fig. 8.4 is an inertial one only in specific cases such as the wind tunnel test of a delta tail fin shown in Fig. 8.5.

All angles are assumed to be small, the drag is neglected, and the tail fin's lift is linear with slope K. Taking moments about the yaw axis and assuming no frictional moment in the yaw bearing results in the following second order linear differential equation:

$$\frac{d^2\theta}{dt^2} + 2\zeta\omega_n\frac{d\theta}{dt} + \omega_n^2\theta = \omega_n^2\varphi \tag{8.4}$$

where φ is the wind direction. The natural frequency ω_n and damping ratio ζ are given by (8.5a, b) respectively:

$$\omega_n = U_{\text{wake}}\sqrt{\frac{\rho rAK}{2I}} \tag{8.5a}$$

and

$$\zeta = \sqrt{\frac{\rho r^3AK}{8I}} \tag{8.5b}$$

where ρ is the air density, A the tail fin area, and I the inertia about the yaw axis. It is important to note that I has contributions from the tail boom, nacelle, and blades. These "pseudo-static" relationships (8.5a, b) were originally derived for wind vanes used to measure wind direction, e.g. Weiringa [9] and Kristensen [10],

Fig. 8.5 Wind tunnel testing of a 1/4 scale model of the tail fin in Fig. 6.1. The vertical shaft has an encoder to measure angle. From Wright [21]

and have been used to model a wind turbine tail fin by Ackerman [11], and Davis and Hansen [12]. For a delta wing tail fin, K is given by

$$K = \pi b/c \tag{8.6}$$

from slender body theory, see Eq. 8.94 of Katz and Plotkin [13]. Note also that $K = K_p$ from Eq. 8.2 for any planform whose lift and drag are describable by the Polhamus equations.

A linear second order equation for yaw performance can also be derived by applying "unsteady slender body" theory (USB), Ebert and Wood [3]; the basic material is in Chap. 13 of Katz and Plotkin [13]. Though still only valid for small angles with drag neglected, this is, in principle, a more comprehensive treatment, as it includes the time-variation of the trailing vorticity and does not ignore the dynamic issues of a non-inertial co-ordinate system. The equation for the moment about the turbine's yaw axis is

$$I\ddot{\theta} = K_1(\theta - \varphi) + K_2\dot{\theta} - K_2'\dot{\varphi} + K_3\ddot{\theta} \tag{8.7}$$

where θ is now the angle between the fin and an "earth-fixed" inertial system whose actual origin is not important as attention is focused on yaw rates. φ is the wind direction in the same co-ordinates. As before, I is the total yaw moment of inertia of the turbine, and

$$\begin{aligned} K_1 &= \frac{1}{4}\pi\rho b^2 U^2 (2/3c + x) = \frac{1}{4}\pi\rho b^2 U^2 r, \\ K_2 &= \frac{1}{4}\pi\rho b^2 U(c+x)^2, \quad K_2' = \frac{1}{4}\pi\rho b^2 U(c^2/4 + cx/3), \quad \text{and} \\ K_3 &= \frac{1}{4}\pi\rho b^2 c(c^2/5 + x^2/3 + cx/2), \end{aligned} \tag{8.8}$$

where the tail boom length $x = r - 2c/3$. The term involving K_1 is the steady lift and is equivalent to (8.6) for the lift slope with the centre of pressure being $2c/3$ from the apex and distance r from the yaw axis. K_2 and K_2' are due to the instantaneous downwash—the trajectory of the flow caused by the lift-produced vorticity—and represents the main change from the pseudo-static equation. A term in the time rate change of wind speed should also appear in (8.7) but many applications of the equation to actual data as described below, suggest that it is not important.

Neither theory is entirely satisfactory: the quasi-static method ignores the unsteadiness of the tail fin wake and the "added mass"—the effective mass of air that moves with the fin—although the analysis presented below shows this error to be small. USB has an added mass term and the unfortunate failing of predicting no steady lift for a rectangular wing, because USB forces are generated by the chordwise changes to the wing shape. Readers may be interested to know that USB has been applied to the aerodynamics of bird tails [14–16].

Figure 8.2 suggests that x is considerably greater than c for most turbines, which, combined with Eq. 8.8 for K_1 implies that the chord of the tail fin only has a small effect on its dynamics. The K_3 term in (8.7) and (8.8) represents the added mass of the air moved with the yawing tail fin. For the case where the wind direction does not change, ω_n and ζ for a delta wing tail fin are:

$$\omega_n = \sqrt{\frac{K_1}{I + K_3}} \approx \sqrt{\frac{K_1}{I}} \tag{8.9a}$$

and

$$\zeta = \frac{K_2}{2\sqrt{K_1(I + K_3)}} \approx \frac{K_2}{2\sqrt{K_1 I}} \tag{8.9b}$$

The approximations in (8.9a, b) are valid when the added mass can be ignored. Note that the natural frequency is proportional to the wind speed, but the damping ratio is determined only by the geometry and inertia.

A wind tunnel test of a one-quarter model tail fin of the 500 W turbine in Fig. 6.1 is shown in Fig. 8.5. The model was mounted to the wind tunnel working section floor. There was also a ceiling that is not visible, but no side walls. At the top left can be seen the inside of the wind tunnel contraction. The wind direction is parallel to the model's tail arm. In other words, the tail fin was not yawed when the photograph was taken. U, the wind speed, was kept constant while the fin was released from a number of different yaw angles. The subsequent yaw angle was measured as a function of time. Figure 8.6 compares a typical result to the response predicted by both methods. Note that the release angle of 40° should be large enough to invalidate the small angle approximation, but repeated experiments over a range of release angles failed to detect any nonlinearities in behaviour due to large angles. For example, the measured period in Fig. 8.6 does not alter significantly as the yaw reduces and the small angle approximation should become valid. The pseudo-static method under-predicts the damping ratio, while USB slightly over-predicts it. Both over-estimate the natural frequency, by an

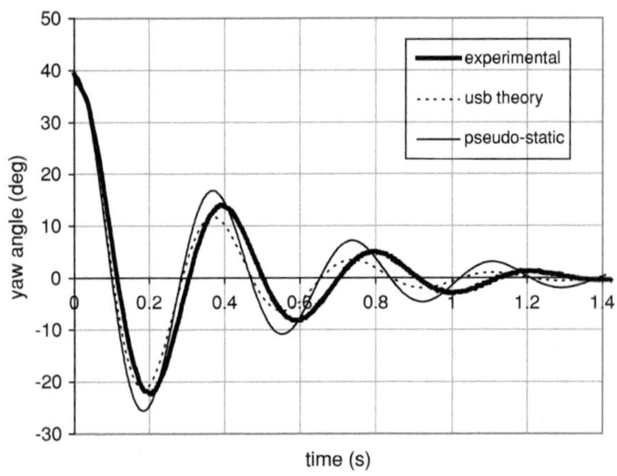

Fig. 8.6 Response of the model tail fin in 15 m/s wind tunnel testing for a release angle of 40° [21]

equal amount, as can be seen by both predictions crossing the x-axis at the same time. This is because the added mass was negligible as explained below.

However, the wind tunnel tests involve simplifications that rarely occur in practice, so that accurately predicting yaw behaviour of an operating turbine is more complex than suggested by Fig. 8.6. For example, the linear second order yaw behaviour of a delta wing is strictly valid only for a constant wind speed, but as mentioned above, it was not possible in practice to determine the effects of non-constant U. The actual tail fin was tested with the rotor immobilized (to remove complications to the yaw behaviour arising from power extraction), with the air density $\rho = 1.18$ kg/m^3. Table 8.2 gives the other relevant parameters.

Before considering the yaw behaviour, it is worth pointing out that the operational Re range of this tail fin, from the blades being stationary at $U_0 = 3$ m/s to rated power at 10 m/s, is approximately 2–6 $\times$ 10^5, which is much less than the range in Re of the operating blades.

Example 8.1 Determine the relationships between the quasi-static and USB natural frequency and damping ratio for a delta wing obeying (8.6) when the added mass can be ignored.

Answer Substituting $A = bc/2$ for a delta wing and (8.6) in (8.5a) gives the approximate version of (8.9a). Thus both theories predict the same natural frequency. Making similar substitutions in (8.5b) and using (8.9b), the ratio of damping ratios is

$$\zeta_{\mathrm{USB}}/\zeta_{\mathrm{QS}} = (1 + c/(3r))^2$$

so that USB always predicts higher damping, but the difference reduces with increasing tail arm length.

Example 8.2 Determine the USB and quasi-static natural frequency and damping ratio for the tail fin detailed in Table 8.2.

Answer Using the tabulated values, $K_1/U^2 = \frac{1}{4} \times \pi \times 1.18 \times 0.44^2 \times 0.96 = 0.1722$ kg, from Eq. 8.8. Similarly $K_2/U = \frac{1}{4} \times \pi \times 1.18 \times 0.44^2 \times (0.62 + 0.51) = 0.2291$ kg m, and $K_3 = \frac{1}{4} \times \pi \times 1.18 \times 0.44^2 \times 0.51 \times (0.51^2/5 + 0.62^2/3 + 0.51 \times 0.62/2) = 0.0311$ kg m^2. Thus the added mass is negligible in comparison to the inertia of the turbine. From Eq. 8.9a, $\omega_n/U \approx \sqrt{}(0.1772/$

Table 8.2 Parameters for tail fin test data in Fig. 8.7

Parameter	Value
$r = x + 2c/3$	0.96 m
A (area of tail fin)	0.112 m^2
x (tail boom length)	0.62 m
c (tail fin chord)	0.51 m
b (tail fin span)	0.44 m
I (moment of inertia of tail fin and boom)	2.74 kg m^2
K (slope of linear portion of lift curve)	2.71/rad from Eq. 8.6

2.74) = 0.251 m^{-1} (note that $U_{\text{wake}} = U$ for these tests), and $\zeta \approx 0.2027/(2\surd(0.1722 \times 2.74)) = 0.166$ from Eq. 8.9b.

Equation 8.5a, gives

$$\omega_n/U = \surd(1.18 \times 0.96 \times 0.112 \times \pi \times 0.44/(2 \times 0.51 \times 2.74)) = 0.251\ \text{m}^{-1},$$

which is equal to that from USB as expected. From (8.5b)

$$\zeta = \surd\left(1.18 \times 0.96^3 \times 0.112 \times \pi \times 0.44/(8 \times 0.51 \times 2.74)\right) = 0.120$$

Note that $0.120 \times (1 + 0.51/(3 \times 0.96))^2 = 1.66$ as required form the results of the previous Exercise.

Only the USB predictions from Eqs. 8.7 and 8.8 are shown in Fig. 8.7 because the significantly higher damping from USB gives a better approximation to the measured yaw response. This difference in damping is significant because the hardest part of tail fin design is to provide sufficiently high damping. As is clear from the Exercises 8.1 and 8.2, the agreement in natural frequency is due to the small magnitude of the added mass term, K_3, in comparison to the inertia. The tail boom contributed 0.747 kg m^2 to I, and the steel tail fin 0.758 kg m^2. The remaining contributions to I come from the rotor, generator, and a lead weight added to the tail to increase the inertia.

The simulation in Fig. 8.7 is fairly accurate and the tail fin has obviously prevented the turbine from responding to the high frequency wind direction changes, but it is reasonable to infer from the sequence 0–5 s, and from 15 to 25 s, that the damping ratio of the tail fin is too low.

Example 8.3 How could the damping ratio of the tail fin described in Table 8.2 be increased?

Answer The USB Eq. 8.9b shows that it is necessary to increase K_2 without increasing K_1 or I. This is not easy. The only parameter in K_2 that is absent from K_1 is the chord, c. Increasing c will increase the damping, but the effect may not be

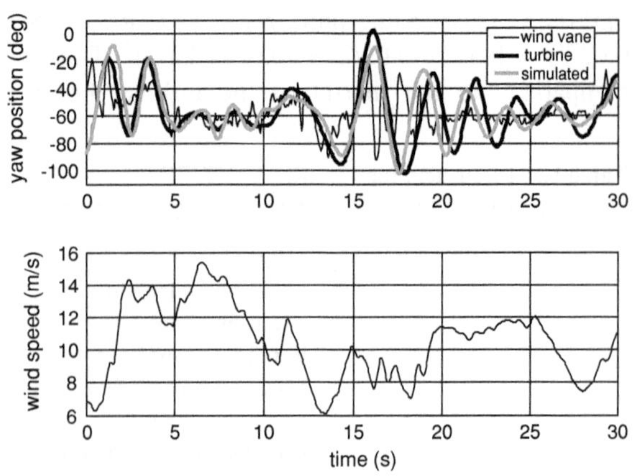

Fig. 8.7 Measured and calculated turbine yaw $\omega_n/U = 0.25$, $\zeta = 0.166$ with stationary blades, from Wright [21]

significant if $x \gg c$, and will tend to be negated by the increase in inertia. In the next section, the effects of changing planforms will be discussed and shown not to significantly alter the lift slope. In other words, there is little the turbine designer can do to increase the damping ratio of conventional tail fins. On the other hand, the twin-delta tail fin studied by Ebert and Wood [3] had a damping ratio of around 0.8.

Yaw performance with the rotor stationary is extremely important in practice. It was pointed out in Chap. 6 that wind direction changes are often greater at low wind speeds where the turbine spends more time starting, so the ability of the tail fin to provide yaw on its own is critical. Yaw behaviour when the turbine is producing power is discussed in Sect. 8.5, where it is shown that the rotor provides yaw stability and the tail fin is less critical.

8.4 Planform Effects on Tail Fin Performance

The analysis of tail fin behaviour started with delta wings because their lift and drag, and centre of pressure are well-known. However, the turbine designer may wish to use a more pleasing shape for the fin or one with a higher lift slope. Unfortunately, there is less information available for planforms other than delta wings. One way to discuss the yaw behaviour of any fin is to assume that it is characterised by the "generalised" natural frequency and damping ratio, found from USB by neglecting the added mass and assuming that K_1 can be expressed as

$$K_1 = \frac{1}{2}\rho AKU^2 r' \tag{8.10}$$

where $r' = x + x_{cp}$ is the distance from the yaw axis to the centre of pressure, located x_{cp} from the leading edge. Thus

$$\omega_n = U_{\text{wake}}\sqrt{\frac{\rho AKr'}{2I}} \tag{8.11a}$$

and

$$\zeta = (c+x)^2\sqrt{\frac{\rho AK}{8Ir'}} \tag{8.11b}$$

and the $K_2{}'$ term in (8.7) can be similarly modified if predictions of turbine yaw behaviour are required. Polhamus [17] computed the behaviour of "arrowhead" and "diamond" planforms which are modifications of a standard delta as shown in Fig. 8.8. The indentation i is taken as positive for diamond and negative for an arrowhead. The aspect ratio is unaffected by any indentation (Fig. 8.8).

Polhamus's results can be fitted by a relationship of the form

$$K = (AR)[K_{P0} + (AR)K_{P1}] \tag{8.12}$$

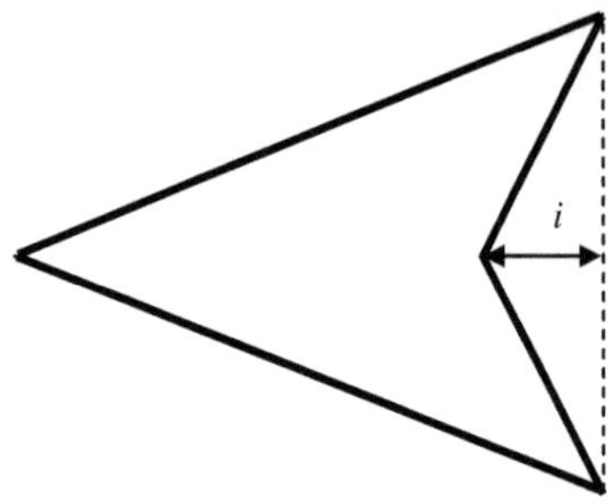

Fig. 8.8 Modification to delta planform to make an *arrow* (as shown) or diamond

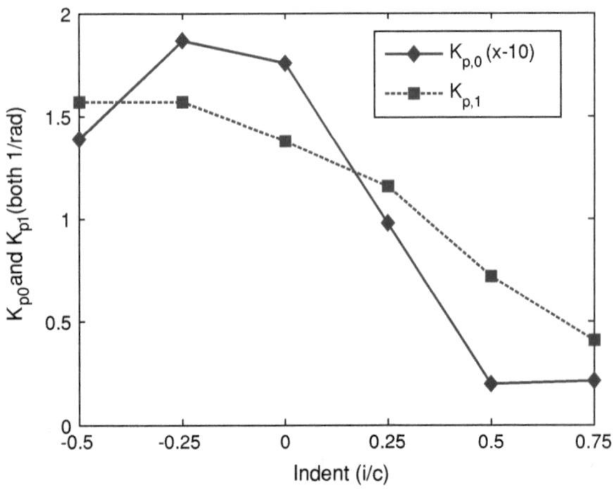

Fig. 8.9 The constants in Eq. 8.12 as a function of indentation

where the constants K_{P0} and K_{P1} are shown in Fig. 8.9 as a function of the indentation. Unfortunately the centre of pressure was not reported and no experimental data appears to be available for it or the lift and drag.

Torres and Mueller [18] measured the lift and drag of a number of three-dimensional shapes, such as rectangles, (but not delta wings) of interest for tail fin design. They correlated their lift and drag data using

$$C_L = K_p \sin\alpha \cos^2\alpha + \pi \cos\alpha \sin^2\alpha, \quad \text{and} \quad C_D = C_{D,0} + kC_L^2 \tag{8.13}$$

The drag equation is of the same form as Eq. 4.13 and the viscous term $C_{D,0}$ is likely to be small in general. The measurements covered the range $0.50 \le AR \le 2$ at $Re = 10^5$. The earlier normal force measurements for rectangular flat plates by Winter for $0.033 \le AR \le 2$ at $Re = 0.3–0.7 \times 10^5$ are summarised in Fig. 15.52 of Katz and Plotkin [13]. Fitting the results of Torres and Mueller [18] for $1.0 \le AR \le 2.0$ (the most appropriate range for tail fins) from their Fig. 10 gives

$$K_p \approx 0.98(AR) + 0.62 \tag{8.14}$$

which implies that the steady lift on a rectangular fin of $AR \approx 1$ (nearly a square) approximates that on a delta fin of the same AR. Note, however, that the delta wing will contribute less to the yaw inertia than will a rectangular fin. The parameter k in the drag equation varied from about 0.7 at $AR = 0.5$ to 0.4 at $AR = 2$. There is little information on the *Re*-dependence of lift and drag for low *AR* flat plates. At low α, the centre of pressure on a rectangular wing is approximately $0.2c$. It is reasonable to assume from these results that tail fin shape has only a second order effect on tail fin performance. Nevertheless, it would be highly desirable to have lift, drag, and centre of pressure data for more possible tail fin shapes, and to explore their yaw behaviour in experiments like that shown in Fig. 8.5.

Example 8.4 In terms of increasing the damping ratio of the tail fin, is it better to use an arrow or a diamond shape rather than a delta wing?

Attempted Answer The data fit in Fig. 8.9 gives information on the variation in K_1 in Eq. 8.8 with shape, but give nothing on K_2. Negative indentation increases the lift slope K_1, which, apparently paradoxically, decreases the damping by Eq. 8.11b, unless the associated reduction in I counteracts this. A similar cancellation occurs the other way; a positive indentation reduces K_1 but increases I. At this point, the best advice is to set up a wind tunnel test similar to that shown in Fig. 8.5 and measure the response of the new shape. Fortunately these tests are straightforward.

8.5 Rotor Effects on Yaw Performance

The contribution to yaw moments from the aerodynamic forces on the rotor is difficult to determine, and a satisfactory understanding does not yet exist. The thrust and the centre of thrust during yaw determine the moment on the rotor. Standard blade element calculations suggest that in the absence of azimuthal variations in inflow, no yaw moment is generated by an unconed rotor, that is a rotor whose blades are not bent in the wind direction, either by the aerodynamic loads or by design. Coning the blades tends to stabilise the rotor in yaw much like dihedral gives roll stability to aircraft wings [19]. Unpublished BET calculations at the University of Newcastle assuming azimuthally-uniform inflow gave a steady restoring moment on a coned 5 kW rotor that was linear in yaw angle below 60°. This is an interesting result, because it implies the linear second order yaw behaviour also applies to a turbine when producing power. Coning can occur either by the downwind deflection of flexible blades under load, or by pre-setting. By combining the BET results with USB, the main change due to coning is to increase the damping which is in agreement with the measurements of Bechly et al. [4]. Other influences on rotor yaw moment include the following [20]:

- Dynamic stall—hysteresis in the aerodynamic performance of the blades (particularly in stall) as they rotate
- Skewed wake induction effects
- Vertical wind component or rotor tilt
- Vertical and horizontal wind shear
- Turbulence—a combination of the above wind conditions varying randomly in time

Typical yaw data when the turbine in Fig. 6.1 is producing power is shown in Fig. 8.10—Wright [21] shows much more data for various tail fin shapes and a range of other parameters. Generally, the simulations (even including an extra linear term for the coning) are less accurate when the turbine is producing power, which is hardly surprising in view of the list immediately above (Fig. 8.10).

In discussing the data in Fig. 8.7, it was suggested that the damping ratio of the tail fin whose details are in Table 8.1, was too low. Examples 8.4 and 8.6 show that it is difficult to alter the damping ratio by changing the planform. The one possibility that has not yet been considered is to alter the tail boom length.

Example 8.5 Can changing the tail boom length influence the damping?

Attempted Answer Equation 8.11b gives $\zeta \sim b(r + c/3)^2/(rI)^{1/2}$ if A and K remain constant. Keeping the tail fin unchanged but increasing r can only be of limited help as for $r/c \gg 1, I \sim r^3$ (see Exercise 8.6), and ζ becomes independent of r.

8.6 High Yaw Rates

Equation 8.1 shows that gyroscopic loads in the rotor shaft and blade root are proportional to the product of rotor speed and yaw rate. This suggests that one way to summarise a large amount of yaw data is to plot yaw rate against rotor speed and wind speed as done in Fig. 8.11 for the 500 W turbine. Chapter 9 discusses the IEC SLM which specifies a maximum yaw rate dependent on rotor area. For the 500 W turbine, this is ±2.99 rad/s = ±171°/s which is assumed to occur at the design Ω. The SLM maximum gyroscopic moment depends on the product of these two angular speeds. Figure 8.11 shows positive and negative extreme yaw rates in excess of this value.

Up to wind speed of around 10 m/s, the maximum yaw rate increases almost linearly with wind speed, possibly as a consequence of the linear relationship

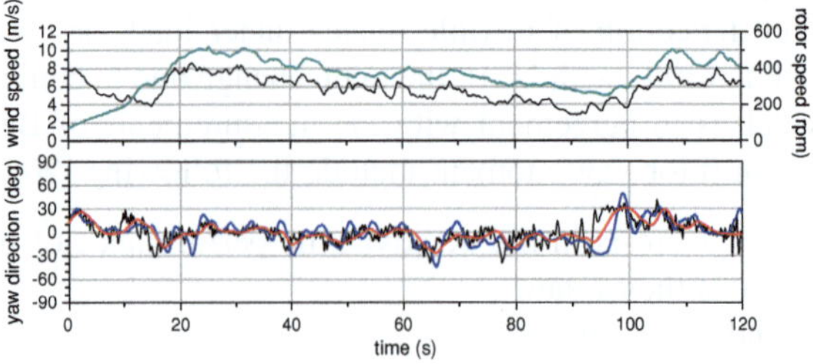

Fig. 8.10 Comparison of computed (*thick red line*) and actual (*thick blue line* with more variations than the computed) yaw behaviour when turbine is producing power. The *thin black line* is the wind direction. On the *upper part* of the figure the rotor speed is above the wind speed (which starts at nearly 8 m/s) for most of the record. Figure from Wright [21]

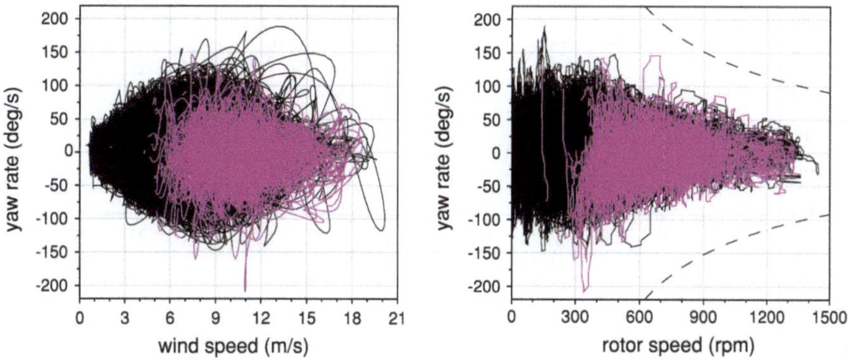

Fig. 8.11 Yaw rate as a function of wind speed and rotor speed, 97 h of data [21]. The pink data indicate the turbine is furling. The *dashed curves* on the right hand give the safety limit on yaw rate explained in the text

between ω_n and U_{wake} in Eqs. 8.5a and 8.9a. However, it is also possible that changes in wind direction decrease in magnitude as wind speed increases and this decrease takes over from the natural frequency in setting the maximum yaw rate at around 10 m/s. However, it has been observed that the peak yaw rates for a given wind speed are typically associated with yaw angles in excess of 40°, outside the range of validity of the theories presented in this chapter. Figure 8.12 is an example of such a high-yaw-rate event, in which a large increase in wind speed follows a change in wind direction. The turbine does not initially respond to the change in wind direction possibly because of the low wind speed. As the wind speed increases, high yaw rates occur during the subsequent realignment of the turbine. At time just after 6 s, the yaw rate exceeds the IEC stipulated maximum. Fortunately, the rotor speed remains low during this high yaw rate, so the gyroscopic loads are not excessive. The dashed lines in the right side of Fig. 8.11 indicate the IEC limits of rotor speed and yaw rate given by Eqs. 9.6 and 9.14b of the next chapter which are based on (8.1). If the limits are exceeded, then the actual gyroscopic moment exceeds that stipulated in the standard. This turbine appears to be performing safely but further tests showed that reducing the moment of inertia of the turbine about the yaw axis caused higher yaw rates.

8.7 Aerodynamic Over-speed Protection

There are two main ways of enabling over-speed protection through the action of aerodynamic forces, namely furling of the tail fin and pitching. They are now considered in turn.

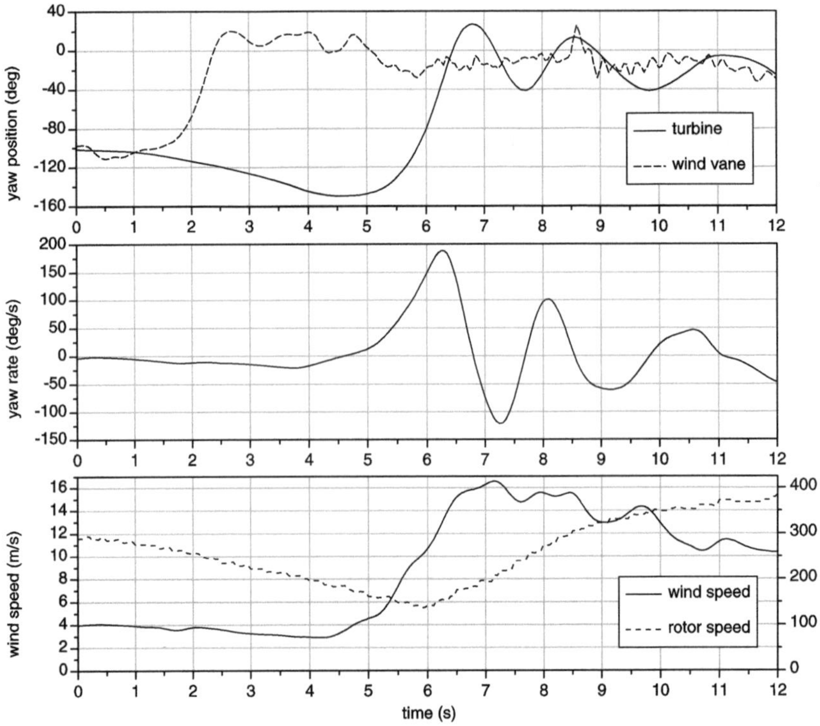

Fig. 8.12 A large wind direction change coupled with a large gust for the turbine in Fig. 6.1 from Wright [21]

8.7.1 Furling

Furling to prevent rotors from over-speeding has been used since before the twentieth century in the old "American" high solidity water pumpers which had no electrical system, e.g. Kentfield [22]. Furling is cheap and easy to implement.

A common furling system is illustrated in Figs. 8.13 and 8.14 for the 500 W turbine from Fig. 6.1. The yaw axis is offset from the rotor axis by a small distance, and the tail fin boom is fixed to the rear of the nacelle by a hinge that is tilted at angle δ (10° in this case) from vertical. This hinge is the furl axis shown as the solid white line in Fig. 8.14. In low winds, the thrust on the rotor acting on the yaw axis offset causes the turbine to operate at a constant, hopefully small yaw angle, balanced by the restoring tail fin moment. The tail is held in place about the furl axis by its own mass. As the wind speed increases, both the rotor thrust moment and the tail fin moment increase. Eventually the gravitational moment of the tail fin about the furl axis is exceeded by the aerodynamic moment on the tail fin, and the tail begins to collapse or furl behind the nacelle. This turns the rotor out of the wind direction, thus reducing its speed and power output.

Fig. 8.13 Rear view of the 500 W turbine showing tail fin. Photo from Wright [21]

Equation 8.15 below expresses the equilibrium between the gravitational and aerodynamic moments of the tail fin at furl axis tilt angle ψ, where m_{tf} is the total mass of the tail fin assembly, r_{cm} is the distance from the furl axis to the centre of mass of the tail fin assembly, r_{cp} is the distance from the furl axis to the fin centre of pressure, and F is the aerodynamic normal force acting on the fin. Note that in its unfurled position at $\psi = 0$, the tail fin is constrained by a rubber stopper.

$$m_{tf} g\, r_{cm} \cos\psi \,\sin\,\delta - F\, r_{cp}\, \cos\delta = 0 \tag{8.15}$$

where δ is the furl axis offset angle defined in Fig. 8.14.

A much more complete analysis of furling is given by Bikdash et al. [23] and other measurements and analysis are given by Jonkman and Hansen [24]. In practice, furling has a number of drawbacks. In marginal furling conditions or gusty winds the tail fin may continually furl and unfurl and so not shut down the turbine. There is evidence that high yaw rates occur during either furling or unfurling as shown by the data in Fig. 8.15 for the furling tail fin of Fig. 8.14. Note that the bottom graph shows the furl angle between the tail arm and the rotor axis: the highest yaw rates occur during furling and unfurling and are associated with higher Ω than in Fig. 8.11. It is also very difficult to design a furling system to act in a particular way from theoretical calculations, so some trial and error is usually required. This problem arises partly from the need for furling to protect against two quite different situations. Furling is routinely required when the turbine is producing full power and the wind speed increases, where from Chaps. 2 and 5, the thrust coefficient is close to unity. Much less often, the turbine will lose its load and the blades accelerate to the runaway state where the thrust coefficient may be significantly higher, judging from Fig. 2.2.

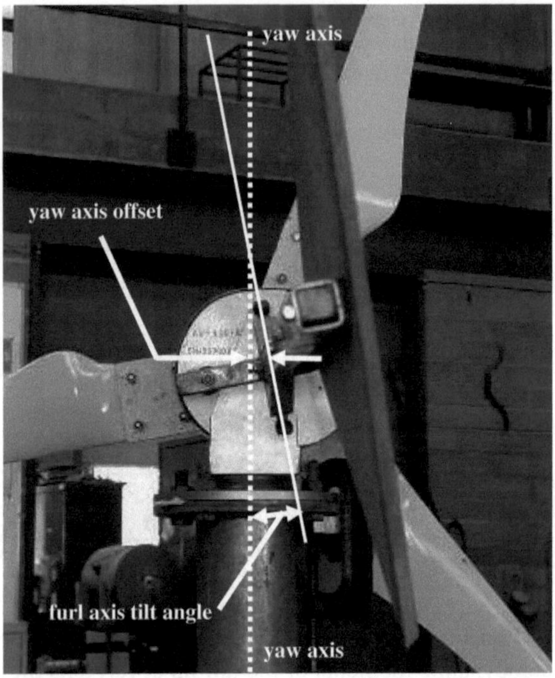

Fig. 8.14 Close up of 500 W turbine furling mechanism. Photo from Wright [21]

Figure 8.16 shows that a yaw axis offset causes a non-zero yaw during normal operation of the 500 W turbine. Each data point represents 15+ seconds of relatively constant wind speed, and a corresponding average yaw angle. When the turbine generates power, the thrust offset causes an average yaw angle of about 20°, which changes little over the range of measured wind speed. This reduces the power output by around 10% from the unyawed case. Furthermore, there can be significant hysteresis in furling, whereby the wind speed required for unfurling is considerably lower than that for the original furling. In field measurements of a 10 kW turbine, Bowen et al. [25] found significant reductions in the power output attributable to this effect. They described the measured power curve as having "two separate concentrations of data points"—see their Fig. 9—due to hysteresis.

Despite the drawbacks of furling, and the complexity it adds to analysing yaw behaviour, its widespread use indicates it is one of the simplest over-speed protection methods for small wind turbines.

8.7.2 Pitching

The main part of a pitching turbine (rotor, generator, and tail) is hinged below and behind the centre of mass such that an increasing wind speed will eventually cause the moment due to turbine thrust to exceed the counteracting moment due to

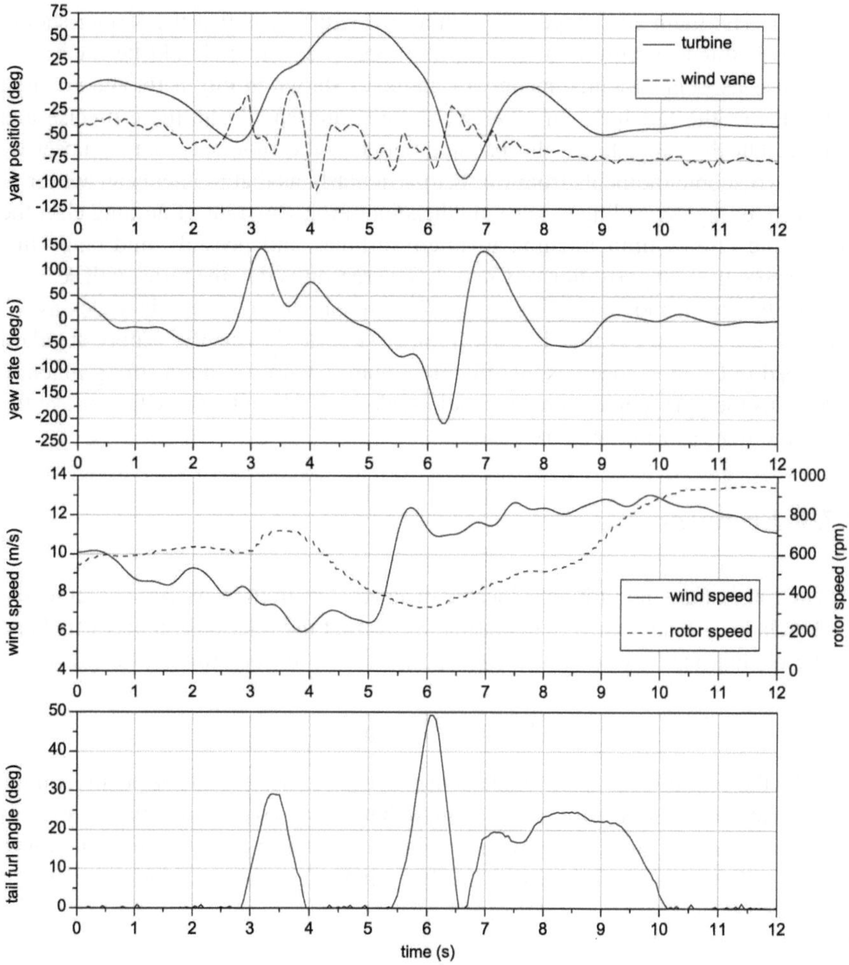

Fig. 8.15 Large yaw rates associated with furling of the 500 W turbine from Wright [21]

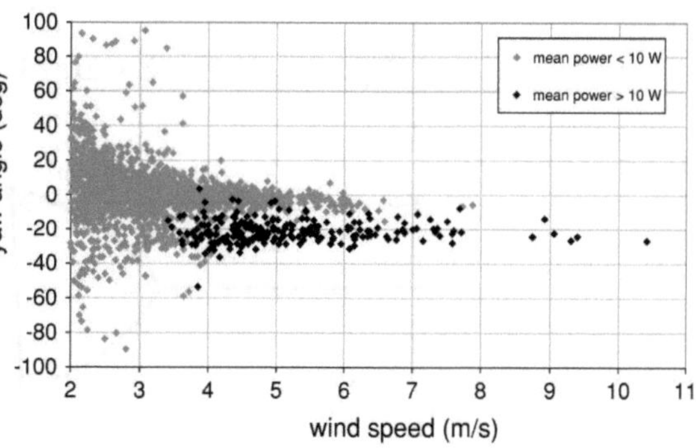

Fig. 8.16 The mean yaw angle during unfurled operation of the furling 500 W turbine [21]

the mass. The main advantages of pitching over furling is that the former should occur over much narrower range of wind speeds, so there should be much less power loss associated with a mean yaw angle, as shown in Fig. 8.16, and no large yaw rates as the tail furls. In other words, pitching should be more precise and definite. On the other hand, the balance between the moments can be upset by strong gyroscopic moments from Eq. 8.1 particularly as that moment can act for or against pitching depending entirely on the sign of the yaw rate. If nothing else, the gyroscopic effects limit pitching to micro-turbines and possibly mid-range turbines, using the categorisation of Chap. 1. The reason is that the thrust will scale as R^2 and its typical moment will scale as R^3 whereas the rotor inertia scales as R^n, where $4 < n < 5$, see Table 1.7, and typical Ω as R^{-1}. The IEC safety standard described in Chap. 9 specifies that the maximum yaw rate is not a strong function of R. Thus the gyroscopic moment should scale as R^{n-1}, so that it will overtake the thrust moment as rotor size increases.

8.7.3 Exercises

1. The rotor of a wind turbine is located distance d from the yaw axis. Assume that the blades are not coned in operation. Show that the relationship between J and the blade's yaw moment of inertia, I_b, as a function of azimuthal angle ψ, is

$$I_b = m_b d^2 + J \sin^2 \psi$$

Take the origin for ψ as the blade being vertical upwards and ψ is positive in the direction of blade rotation. Also assume that $c/r \ll 1$ so that the blade's mass is concentrated in the radial direction. Find the average and peak-to-peak variation in I_b with ψ for rotors with (a) two blades, (b) three blades, and (c) four blades.

2. Derive Eq. 8.4.
3. With a set-up similar to Fig. 8.5, how would you devise an experiment to determine the significance of the added mass for the tail fin response?
4. A delta tail fin of constant thickness t is made from a material of density ρ_{tail}. Show that its mass moment about the yaw axis is

$$I_{\text{tail}} = \rho_{\text{tail}} bct\left[x^2/2 + 2xc/3 + c^2/4\right]$$

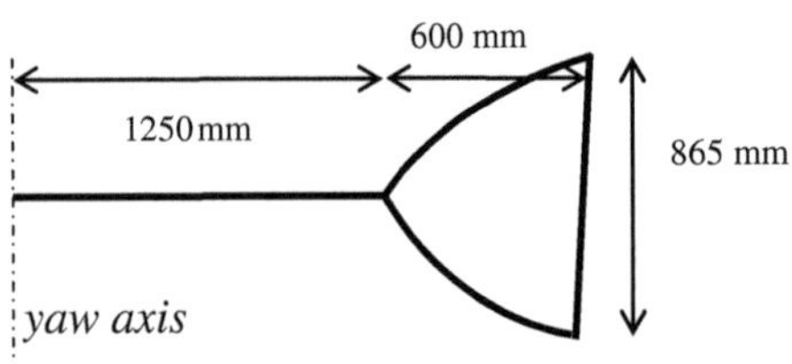

5. The tail fin for the Aerogenesis 5 kW turbine shown in Fig. 1.2 is sketched above. The fin is made from 3 mm steel plate of density 7800 kg/m^3. Estimate the inertia, natural frequency and damping ratio of the fin on its own. Recalculate these values using the total inertia of tail fin, turbine, and blades about the yaw axis of 58.05 kg m^2.
6. If the tail boom has a constant but small cross-sectional area, A_b, then its moment of inertia about the yaw axis, I_{boom}, is $\rho_b A_{\text{boom}} r^3/3$, where ρ_{boom} is the density of the tail boom material. Hence show that if $c/r \ll 1$, where c is the tail fin chord, then the total moment of inertia of the boom and fin is approximately

$$I \approx \rho_b A_b r^3/3 + m_{\text{tail}} r^2$$

7. Determine the limits on the designer's ability to adjust the damping ratio and natural frequency by altering the tail boom length.
8. Assume that the yaw moment of inertia is dominated by that of the rotor and tail fin and that all component lengths scale on tip radius R. How will the damping ratio in yaw change with increasing R? You may wish to review Sect. 1.9 at this point.
9. Do a web search of manufacturers of wind turbines in the range 40–100 kW. Do any of these have tail fins? Comment on the answer to this question.

10. The picture shows the Vergnet 7–10 small wind turbine. The "7" refers to the blade diameter in m and "10" to the rated power output in kW. The very distinctive design ensures that the tail fin is not in the wake of the nacelle and tower. Assuming that the tail boom angle is 45°, estimate the difference in natural frequency and damping ratio when compared a similar tail fin on a horizontal boom of the same length.
11. Using the results of the previous exercise, would you use a rectangular tail fin in preference to a delta wing of the same area? Assume the tail fin chord in each case is much less than the tail boom.
12. It has been suggested that one way to get extra damping when needed is to have the fin "spring-loaded" so it can slide along the tail boom and move further away from the yaw axis as the centripetal force due to high yaw rates overcome the spring tension. Comment on this suggestion.
13. Is it possible to design a furling tail fin that does not have a "yaw error" when unfurled?

References

1. Leishman JG (2002) Challenges in modeling the unsteady aerodynamics of wind turbines. Wind Energy 5:86–132
2. Leishman JG (2006) Principles of helicopter aerodynamics, 2nd edn. CUP, Cambridge
3. Ebert PR, Wood DH (1995) On the dynamics of tail fins and wind vanes. J Wind Eng Ind Aerodyn 56:137–158
4. Bechly ME, Gutierrez H, Streiner S, Wood DH (2002) Modelling the yaw behaviour of small wind turbines. Wind Eng 26:223–239
5. Wright AD, Wood DH (2004) The starting and low wind speed behaviour of a small horizontal axis wind turbine. J Wind Eng Ind Aerodyn 92:1265–1279
6. Polhamus EC (1966) A concept of the vortex lift of sharp-edge delta wings based on a leading-edge suction analogy. NASA TN D-3767, December 1966
7. Polhamus EC (1971) Predictions of vortex-lift characteristics by a leading-edge suction analogy. J Aircraft 8:193–199
8. Bertin J, Cummings R (2009) Aerodynamics for engineers, 5th edn. Pearson Prentice Hall, Upper Saddle River
9. Weiringa J (1967) Evaluation and design of wind vanes. J Appl Meteorol 6:1114–1122
10. Kristensen L (1994) Cups, props and vanes. Riso National Laboratory, Roskilde. http://www.risoe.dk/rispubl/VEA/ris-r-766.htm. (accessed 21 Feb 2010)
11. Ackerman MC (1992) Yaw modelling of small wind turbines. J Wind Eng Ind Aerodyn 39:1–9
12. Davis D, Hansen C (2002) Operation and load measurements during extreme wind events for a southwest windpower whisper H40. AWEA Windpower '02, Portland, Oregon
13. Katz J, Plotkin A (2001) Low speed aerodynamics, 2nd edn. Cambridge University Press, Cambridge
14. Thomas ALR (1993) On the aerodynamics of birds tails. Philos Trans R Soc Biol Sci 340:361–380
15. Thomas ALR (1997) On the tails of birds. What are the aerodynamic functions of birds' tails, with their incredible diversity of form? Bioscience 47:215–225

16. Evans MR (2010) Birds' tails do act like delta wings but delta-wing theory does not always predict the forces they generate. Proc R Soc Biol Sci 270:1379–1385
17. Polhamus EC (1971b) Charts for predicting the subsonic vortex-lift characteristics of arrow, delta, and diamond wings. NASA technical note D-6243
18. Torres GE, Mueller TJ (2004) Low aspect ratio aerodynamics at low Reynolds numbers. AIAA J 42:865–873
19. Miller RH (1979) On the weathervaning of wind turbines. J Energy 4:319–320
20. Hansen AC (1992) Yaw dynamics of horizontal axis wind turbines: final report NREL technical report TP 442-4822. http://www.nrel.gov/docs/legosti/old/4822.pdf. (accessed 29 Nov 2010)
21. Wright AK (2005) Aspects of the aerodynamics and operation of a small horizontal axis wind turbine. PhD thesis, School of Engineering, University of Newcastle
22. Kentfield JAC (1996) The fundamentals of wind-driven water pumpers. Gordon & Breach, Amsterdam
23. Bikdash M, Chen DA, Harb M (2001) A hybrid model of a small autofurling wind turbine. J Vib Control 7:127–148
24. Jonkman JM, Hansen AC (2004) Development and validation of an aeroelastic model of a small furling wind turbine. NREL/CP-500-39589. http://www.nrel.gov/docs/fy05osti/39589.pdf. (accessed 17 May 2010)
25. Bowen AJ, Zakay N, Ives RL (2003) The field performance of a remote 10 kW wind turbine. Renew Energy 28:13–33
26. Kegelman JT, Roos FW (1989) Effects of leading-edge shape and vortex burst on the flowfield of a 70-degree-sweep delta wing. AIAA Paper #89-0086
27. Traub LW, Moeller B, Rediniotis O (1998) Low-Reynolds-number effects on delta-wing aerodynamics. J Aircraft 35:653–656
28. Wentz WH, Kohlman DL (1971) Vortex breakdown on slender sharp edged wings. J Aircraft 8:156–161
29. Matsumiya H, Tsutsui Y, Kumeoka Y (1985) Optimum shapes for the tail-wings of wind turbines. Trans Jpn Soc Mech Eng Ser B 51:3312–3316
30. Koenig DG (1959) Low-speed tests of semispan-wing models at angles of attack from 0° to 180°. NASA memorandum 2-27-59A, Ames Research Center, April 1959

Chapter 9
Using the IEC Simple Load Model for Small Wind Turbines

9.1 Introduction

The IEC safety standard for small wind turbines, IEC 61400-2 and the Germanischer Lloyd guidelines, GL [1],[1] provide a major challenge and opportunity for small wind turbine designers. It was mentioned in the first sentence of the Preface that IEC 61400-2 defines a small turbine as having a rotor swept area of less than 200 m^2 which corresponds roughly to a rated power of less than 50 kW. The standard makes a further provision for micro-turbines of rotor area of 2 m^2 or less; they can be certified independently of their tower. For all other turbines, the tower must be assessed with the turbine.

IEC 61400-2 allows the analysis of turbine safety through three mechanisms:

1. The "Simple Load Model" (SLM) which combines straightforward, and possibly simplistic, equations for the main loads with high safety factors. As its name implies, it is the simplest design methodology and is the only one described in detail in this chapter.
2. "Aero-elastic" Modeling involves (hopefully) more accurate computer modeling of wind turbine loads in response to stochastic inputs such as changes in wind direction and gusts. This method is routinely used for large wind turbines, employing programs such as bladed, which is sold by the consultants Garrad Hassan.[2] The equations for gusts and wind direction changes that are specified in IEC 61400-2 for aero-elastic modeling are identical to those used for large wind turbines, IEC 61400-1, and described by Burton et al. [2]. The main problem with aero-elastic modeling would seem to be the high cost of the software and/or the large amount of time required for its implementation. Aero-elastic modeling is not commonly used for small wind turbines.

[1] GL [1] is the first edition of the guidelines with specific information on small turbines.

[2] http://www.gl-garradhassan.com/en/GHBladed.php, an educational version is available.

D. Wood, *Small Wind Turbines*, Green Energy and Technology,
DOI: 10.1007/978-1-84996-175-2_9, © Springer-Verlag London Limited 2011

3. Load measurements with extrapolations for extreme conditions. Obviously this requires field testing, whereas the first two methodologies can be used at the design stage.

The SLM is unique to the small wind turbine standard IEC 61400-2. It was developed as a cheap alternative to the complex and expensive aero-elastic modeling that is routinely done for large turbines. The "cost" of the SLM is the high safety factors.

This chapter describes an assessment of the three-bladed 500 W, 2 m diameter turbine, shown in Fig. 6.1 and discussed extensively in that chapter and Chap. 8. Although the rotor swept area is larger than 2 m^2, the tower is not considered in detail because a special purpose tower was required for roof installation, that would not be suitable for normal installation. In addition, tower design, which must satisfy standards additional to IEC 61400-2, is discussed in detail in Chap. 10.

Much of the notation used in IEC 61400-2 is also used in this book, but there are some major differences which are collected in Table 9.1. Many of the equations of this chapter are taken from the Standard. This is indicated by a footnote for each equation giving its number in IEC 61400-2.

The first step in applying the SLM is to decide on the turbine class as defined in Table 9.2. Table 9.3 taken from IEC 61400-2, defines the ten load cases comprising the SLM. Note that only Case A is a fatigue load.

Section 9.2 describes the determination of the loads for the 500 W turbine considered in Chaps. 6 and 8. The following sections show how to combine the loads to give the component stresses and present the partial load factors. Section 9.6 introduces an excel spreadsheet to simplify the analysis of this and any other turbine. The extra testing required for IEC certification is described in Sect. 9.7. The last section contains the concluding remarks. A number of shortcomings and errors in IEC 61400-2 are identified in this chapter. This is done with the aim of stimulating further measurements and analysis of wind turbine loads to

Table 9.1 Differences in notation from IEC 61400-2

Definition	Symbol used here	Symbol used in IEC 61400-2
Moment of inertia (kg m^2)	J	I
Number of blades	N	B
Wind speed (m/s)	U	V
Number of cycles to fatigue	N_{cycles}	N
Yaw rate	ω	ω_{yaw}
Blade angular velocity (rad/s)	Ω	n (in rpm), ω_{n}

Table 9.2 The IEC turbine classes

Class	I	II	III	IV	Special (S)
U_{ref}	50	42.5	37.5	30	Values to be specified
U_{ave}	10	8.5	7.5	6	

Table 9.3 The design load cases of the simple load model

Design situation	Load case	Description	Type of analysis
Power production	A	Normal operation	Fatigue
	B	Yawing	Ultimate
	C	Yaw error	Ultimate
	D	Maximum thrust	Ultimate
Power production plus occurrence of fault	E	Maximum rotational speed	Ultimate
	F	Short at load connection	Ultimate
Shutdown	G	Shutdown (braking)	Ultimate
Parked (idling or standstill)	H	Parked wind loading	Ultimate
Parked at fault conditions	I	Parked wind loading (maximum exposure)	Ultimate
Transport, assembly, maintenance and repair	J	To be stated by manufacturer	Ultimate

increase our currently poor knowledge. In this context, it is noted that the author was the Australian representative on the maintenance team that produced IEC 61400-2 and shares the responsibility for the good and bad features of the standard. This chapter is an expanded version of Wood [3].

9.2 The Simple Load Model

The SLM can be applied only to a horizontal axis turbine having two or more cantilevered blades, and a rigid (non-teetering) hub. All IEC standards exclude vertical axis turbines.

The first step involves the collection or assigning of the following parameters:

- Design rotational speed, Ω_{design}
- Design wind speed, $U_{\text{design}} = 1.4U_{\text{ave}}$ (U_{ave} is the average wind speed)
- Design shaft torque, Q_{design}
- Maximum yaw rate, $\omega_{\text{yaw,max}}$ and
- Maximum rotational speed, $\Omega_{\max}$ (Table 9.4)

The design rpm is set by aerodynamic considerations whereas the maximum is set by the overspeed protection, furling in this case. The design power, P_{design}, is determined at the design wind speed with the nominal electrical load connected. The design torque, Q_{design}, is found using[3]

$$Q_{\text{design}} = \frac{30P_{\text{design}}}{\eta\pi\Omega_{\text{design}}} \tag{9.1}$$

[3] (IEC 51).

Table 9.4 Subscripts and co-ordinates

Subscript	Meaning
ave	Average
B	Blade
cog	Centre of gravity (mass)
design	Input parameter for the simplified design equations
e50	Once per 50 year extreme (averaged over 3 s)
f	Loads (applied only to safety factors)
hub	Hub
m	Material (applied only to safety factors)
max	Maximum
r	Rotor
s	Shaft
x	Blade: horizontal direction giving a positive moment in direction of rotation
x	Shaft: horizontal direction such that a positive moment acts in direction of rotation
y	Blade: horizontal direction such that a positive moment is in direction of rotation
z	Blade: towards blade tip

for P_{design} in Watts. The efficiency, η, is to be taken as[4]

$$\eta = \left\{ \begin{array}{ll} \left(0.6 + 0.005 P_{\text{design}}\right) & \text{for } P_{\text{design}} < 20\,\text{kW} \\ 0.7 & \text{for } P_{\text{design}} > 20\,\text{kW} \end{array} \right\} \tag{9.2}$$

The maximum yaw rate, $\omega_{\text{yaw,max}}$, is determined (in rad/s) according to[5]

$$\omega_{\text{yaw,max}} = \left\{ \begin{array}{ll} 3 - 0.01\left(A_{\text{proj}} - 2\right) & \text{for } A_{\text{proj}} > 2\,\text{m}^2 \\ 3 & \text{for } A_{\text{proj}} \leq 2\,\text{m}^2 \end{array} \right\} \tag{9.3}$$

where the "projected" rotor area, A_{proj}, is in m^2. According to Eq. 9.3, $\omega_{\text{yaw,max}}$ reduces from 3 rad/s for micro-turbines to 1 rad/s for the largest possible small turbine. A major limitation of the correlation that is apparent from Chap. 8 is that it is the same for furling and non-furling turbines. Another shortcoming is that (9.3) gives a yaw rate well in excess of the maximum rate of change of wind direction in the model of the "extreme direction change" (EDC) of the IEC Standards for the aero-elastic analysis. Since Chap. 8 demonstrated that wind turbines in yaw approximate a second order linear system when they are not furling, it is unlikely that the maximum yaw rate can exceed the maximum rate of change in wind direction. Thus one way for the designer to "reduce" the gyroscopic loading on the

---

4 (IEC 50).

5 (IEC 27).

blades and shaft is to use the EDC with an aero-elastic analysis rather than the SLM.

9.2.1 Load Case A: Normal Operation

The first load case covers fatigue behaviour during normal operation of the turbine and is the only fatigue load case. It is implicitly assumed that abnormal operation occurs sufficiently seldom to contribute to the fatigue life. All the equations in this sub-section define peak-to-peak loads generated by the assumed cycling between 1.5 times and 0.5 times the design values of the angular velocity and torque. The first is the centrifugal load[6]:

$$\Delta F_{zB} = 2m_B R_{\text{cog}} \Omega^2 \tag{9.4}$$

where R_{cog} is the radius of the centre of gravity (mass) of the blade. The next two equations give, in order, the lead-lag (in the direction of rotation) and flapwise moment (in the direction of the wind; alternatively, out of the plane of rotation)[7,8]:

$$\Delta M_{xB} = Q_{\text{design}}/N + 2m_B g R_{\text{cog}} \tag{9.5}$$

$$\Delta M_{yB} = \lambda_{\text{design}} Q_{\text{design}}/N \tag{9.6}$$

These moments are to be applied to that part of the blade root with the lowest ultimate strength. Note that most turbine blades are very stiff in the lead-lag direction, so that the use of the term "bending" in conjunction with "moment" can be taken to imply bending in the flapwise direction only.

The next three equations give the peak to peak fatigue loads on the turbine shaft. This load is assumed to occur at the first shaft bearing (nearest to the rotor)[9,10,11]:

$$\Delta F_{\text{x-shaft}} = 3\lambda_{\text{design}} Q_{\text{design}}/(2R) \tag{9.7}$$

$$\Delta M_{\text{x-shaft}} = Q_{\text{design}} + 2m_r g e_r \tag{9.8}$$

$$\Delta M_{\text{shaft}} = 2m_r g L_{rb} + \frac{R}{6} \Delta F_{\text{x-shaft}} \tag{9.9}$$

6 (IEC 21).

7 (IEC 22).

8 (IEC 23).

9 (IEC 24).

10 (IEC 25).

11 (IEC 26).

where m_r is the rotor mass, and the rotor eccentricity, e_r, is to be taken as $0.005R$ unless it can be proven otherwise. e_r was defined and discussed in Sect. 7.7. L_{rb} is the distance from the rotor to the first bearing, usually on the front end of the generator. The loads on the blades and shaft that comprise this Load Case will be transmitted to the tower. They are considered in Chap. 10 on tower design.

9.2.2 Load Case B: Yawing

The equation for the maximum blade root bending moment, M_{yB}, is[12]

$$M_{yB} = m_B \omega_{\max}^2 L_{rt} R_{\text{cog}} + 2\omega_{\max} J_B \Omega + \frac{R}{9} \Delta F_{\text{x-shaft}} \tag{9.10}$$

where L_{rt} is the distance from the rotor to the yaw (tower) axis, the maximum yaw rate is determined from (9.3), and $\Delta F_{\text{x-shaft}}$ comes from (9.7). The first term is the centrifugal term. The second term gives the amplitude of the cyclic gyroscopic load; the actual moment is the amplitude multiplied by the cosine of the azimuthal angle measured from the vertically upward position of the blade. Note that this term provides the "envelope" shown on the measured blade root bending moment in Fig. 8.1. The last term approximates the effect of wind shear. A glaring omission is a term accounting for the blade thrust. For two-bladed turbines, the shaft loading is[13]:

$$M_{\text{shaft}} = 4\omega_{\max} \Omega J_B + m_r g L_{rb} + \frac{R}{6} \Delta F_{\text{x-shaft}} \tag{9.11}$$

For a three or more bladed turbine with N blades[14]:

$$M_{\text{shaft}} = N\omega_{\text{yaw}} \Omega J_B + m_r g L_{rb} + \frac{R}{6} \Delta F_{\text{x-shaft}} \tag{9.12}$$

For $N > 1$, there is no contribution from the average blade thrust to the shaft moment. Experience has shown that the gyroscopic terms involving $\omega_{\max}$ are dominant. The difference between (9.11) and (9.12) implies a significant difference between a two-bladed ($N = 2$) and other multi-bladed turbines—recall that the standard specifically excludes single blade rotors. The difference arises because the rotor's moment of inertia about the yaw axis must depend on the azimuthal position of the blades as considered in Exercise 8.1. This is most easily seen for a two-bladed rotor by considering the yaw inertia for the two extreme cases.

---

12 (IEC 28).

13 (IEC 29).

14 (IEC 30).

When the blades are vertical, the blade inertia, J_B, about its axis contributes almost nothing to the yaw moment of inertia. When the blades are horizontal, blade inertia contributes significantly to yaw inertia. The significant variation in inertia alters the maximum loads on the blade and shaft in the manner reflected by the first term in (9.11). For $N \geq 3$, the inertia varies much less with azimuthal position of the blades and can usually be ignored as suggested by the simple analysis in Exercise 8.1. This statement implies that (9.12) approximates both the average and maximum moment on the shaft of a turbine with three or more blades, whereas (9.11) gives the maximum moment for a two-bladed turbine. It is also worth noting that the gyroscopic moment can give rise to a fatigue load, particularly if the tower has a resonant frequency matching the blade passing frequency or the natural frequency of yaw. There is no specific load case in the SLM covering tail fin fatigue.

The derivation of (9.11) and (9.12) (and all the SLM equations) is given in Annex F of IEC 61400-2. The gyroscopic terms are mentioned very briefly in Burton et al. [2], and treated in much more detail by Eggleston and Stoddard [4]. To complicate matters even further: note that the gyroscopic terms do not include any contributions from the yaw acceleration or the angular acceleration of the blades. The former makes a contribution to M_{yB} that is out of phase with the Coriolis term and hence will not alter its magnitude. The latter contributes only to the lead-lag motion of the blade, that is motion in the direction of the rotation where most blades are very stiff. Current knowledge of the extra terms is too poor to justify any attempt at quantification.

The gyroscopic moments acting on the main shaft will be transmitted to the tower. These loads are discussed in Chap. 10 on tower design.

9.2.3 Load Case C: Yaw Error

This load occurs when the turbine operates with a mean yaw error, as was the case of the furling turbine described in the previous chapter. For simplicity, this load case considers a single yaw error of 30°. Yaw error causes a bending moment on the blades according to[15]:

$$M_{yB} = \frac{1}{8}\rho A_{\text{proj},B} C_{l,\max} R^3 \Omega^2_{\text{design}} \left[1 + \frac{4}{3\lambda_{\text{design}}} + \left(\frac{1}{\lambda_{\text{design}}} \right)^2 \right] \tag{9.13}$$

where $A_{\text{proj},B}$ is the projected area of the blades which can be taken as the planform area, and $C_{l,\max}$ is the maximum lift coefficient. If no value is available assume $C_{l,\max} = 2.0$.

[15] (IEC 31).

9.2.4 Load Case D: Maximum Thrust

This is a problematic load because it increases as the square of the wind speed, so that selecting a sufficiently high speed, will produce very high loads. The compromise adopted in IEC 61400-2 is to use[16]

$$F_{x-\text{shaft}} = 3.125 C_T \rho U_{\text{ave}}^2 \pi R^2 \tag{9.14}$$

where C_T is the thrust coefficient, taken to be 0.5. The factor of 3.125 arises from applying (9.14) at a wind speed $U = 2.5U_{\text{ave}}$, which is claimed to give thrust loads comparable with those obtained from aeroelastic modeling of small turbine blades. The thrust must also be transmitted to the tower, making it, along with the turbine weight, and the gyroscopic loads described above, the main ways in which the turbine loads influence the tower loads.

9.2.5 Load Case E: Maximum Rotational Speed

The centrifugal loading on the blades and especially the blade hub can become extremely large at high rotational velocities. The centrifugal load in the blade root is calculated using[17]:

$$F_{zB} = m_B \Omega_{\max}^2 R_{\text{cog}} \tag{9.15}$$

and the bending moment in the shaft due to small blade unbalances is found using[18]:

$$M_{\text{shaft}} = m_r g L_{rb} + m_r e_r \Omega_{\max}^2 L_{rb} \tag{9.16}$$

9.2.6 Load Case F: Short at Load Connection

If a direct short occurs across the output terminals of the generator, a large moment is created on the turbine shaft as a result of an increase in the generator torque, known as the short circuit torque. The following moments arise from this increase in torque[19]:

---

[16] (IEC 32).

[17] (IEC 33).

[18] (IEC 34).

[19] (IEC 35).

$$M_{\text{x-shaft}} = GQ_{\text{design}} \tag{9.17}$$

and[20]

$$M_{xB} = M_{\text{x-shaft}}/N \tag{9.18}$$

where the numerical factor $G = 2.0$ unless a more accurate value is established. This factor obviously relates the short circuit torque to the turbine's design torque.

Note that the order of treatment of the loads used in cases A to E the blades first and then the shaft—has been reversed in this load case and the next. This is done to follow the order in IEC 61400-2.

9.2.7 Load Case G: Shutdown (Braking)

This load case applies to turbines with some form of mechanical or electrical braking within their drive train. It is necessary to know M_{brake}, the extra moment produced by the braking system, either by testing or calculation. M_{brake} must be multiplied by the gearbox ratio if the braking system acts on the high-speed shaft. The shaft moment is[21]:

$$M_{\text{x-shaft}} = M_{\text{brake}} + Q_{\text{design}} \tag{9.21}$$

If the turbine has a brake and a gearbox, $M_{\text{x-shaft}}$ should be increased to account for the drive train dynamics. If the necessary factor cannot be determined accurately, a value of 2 is to be used.

The blade loading caused by the shutdown is calculated by[22]:

$$M_{xB} = M_{\text{x-shaft}}/N + m_B g R_{\text{cog}} \tag{9.22}$$

where the value of $M_{\text{x-shaft}}$ is determined according to Eq. 9.21.

9.2.8 Load Case H: Parked Wind Loading

This load case considers several separate loads acting on a parked turbine, that is, a turbine whose rotor is not producing power. It is not necessary for the blades to be stationary while parked. The loads are calculated using a wind speed of U_{e50}, the 3s 50-year extreme wind speed. The main loading on a stationary rotor is due to drag[23]:

---

20 (IEC 36).

21 (IEC 37).

22 (IEC 38).

23 (IEC 39).

$$M_{yB} = \frac{1}{4} C_d \rho U_{e50}^2 A_{proj,B} R \tag{9.23}$$

where the drag coefficient C_d is taken as 1.5. If the blades are rotating, M_{yB} is due to the lift created on the blades due to variations in the wind direction[24]:

$$M_{yB} = \frac{1}{6} C_{l,\max} \rho U_{e50}^2 A_{proj,B} R \tag{9.24}$$

If there is no value available for $C_{l,\max}$ then a value of 2 is used. Next is the thrust caused by the wind loading on the blades. For a parked rotor, the analogue of (9.14) is[25]

$$F_{x-shaft} = \frac{1}{2} N C_d \rho U_{e50}^2 A_{proj,B} \tag{9.25}$$

For a spinning rotor[26]

$$F_{x-shaft} = 0.17 N A_{proj,B} \lambda_{e50}^2 \rho\, U_{e50}^2 \tag{9.26a}$$

where[27]

$$\lambda_{e50} = \Omega_{\max} \pi R / (30 U_{e50}) \tag{9.26b}$$

This load case also covers the maximum bending moment on the tower base due to the thrust loading on the turbine calculated above. It must also include the wind load on components such as the nacelle and tower, obtained, for all turbine components, from the same basic equation[28]:

$$F = \frac{1}{2} C_f \rho\, U_{e50}^2 A_{proj} \tag{9.27}$$

where A_{proj} is the perpendicular projected area of the component against the wind and C_f is the force coefficient from Table 9.5. This is the first of only two direct references to the tower. The second, in Sect. 11.1 of IEC 61400-2, requires that "support structures shall also meet local codes and regulations…" This may cause difficulties because most "local codes" use lower partial load safety factors than IEC 61400-2.

---

24 (IEC 40).

25 (IEC 41).

26 (IEC 42).

27 (IEC 43).

28 (IEC 44).

Table 9.5 Force coefficients for Load Case H from IEC 61400-2

Shape	Circle	Octagon	Square	Non-lifting body or aerofoil at 90°	Lifting aerofoil
Characteristic length < 0.1 m	1.3	1.3	1.5	1.5	2.0
Characteristic length > 0.1 m	0.7	1.2	1.5	1.5	2.0

9.2.9 Load Case I: Parked Wind Loading, Maximum Exposure

This load case considers the possibility of a failure in the yaw mechanism, such as a seizure of the main yaw bearing. This could cause the turbine to be exposed to a maximum wind velocity in the turbine's most unfavourable position (i.e. maximum projected area). For this reason, the lift and drag on tail fins at high α was considered in Chap. 8. These load calculations, which must include all affected components, use U_{ref}. The force on each individual component is found using[29]:

$$F = \frac{1}{2} C_f \rho\, U_{\text{ref}}^2 A_{\text{proj}} \tag{9.28}$$

The projected area for aerofoil shapes will be their planform area (C_f may result from either lift or drag effects). C_f is again found from Table 9.5.

9.2.10 Load Case J: Transportation, Assembly, Maintenance and Repair

All loading conditions on the turbine from conditions outside the range of normal operation fall into this Case. It includes:

- Gravitational loads during transportation
- Loads caused by special tools
- Wind loads during installation
- Loads associated with hoisting the turbine into position
- Loads on a tilt up tower during erection
- And loads associated with the support structure from climbing it.

A calculation for the bending moment at the base of the tower due to lifting the turbine into position is given as[30]:

$$M_{\text{tower}} = 2\left(m_{\text{tower top}} + m_{\text{over hang}}/2\right) g\, L_{lt} \tag{9.29}$$

[29] (IEC 45).

[30] (IEC 46).

where the two masses in parentheses are, in order, that of the nacelle and rotor and that between the lifting point and tower top, and L_{lt} is the distance between the lifting point and the tower top. Loads occurring during tower raising and lowering are discussed in Chap. 11 where the importance of only raising and lowering the turbine on calm days is emphasised.

9.3 Stress Calculations and Safety Factors

The loads calculated in Sect. 9.2 must be converted into equivalent component stress levels to compare with the allowable material stress limit. Stress levels are calculated by the same procedure regardless of whether the SLM or aero-elastic analysis was used.

9.3.1 Equivalent Component Stresses

Following the SLM calculations, the individual forces and moments are combined to obtain the final equivalent stress level on the key load carrying components, the main shaft and the blade root using the formulae in Table 9.6. The calculation of the equivalent stress levels has to take into account a number of key factors:

- Stress variations within the component
- Stress concentrations
- The size and direction of the resulting load or stress
- Variations in component dimensions and thickness
- Component surface treatment
- The type of loading on the component
- Any manufacturing effects on the components such as welding, machining etc.

Table 9.6 Equivalent stress formulae from IEC 61400-2

	Circular blade root	Rectangular blade root	Rotor shaft
Axial load	$\sigma_{zB} = \frac{F_{zB}}{A_B}$	$\sigma_{zB} = \frac{F_{zB}}{A_B}$	$\sigma_{\text{x-shaft}} = \frac{F_{\text{x-shaft}}}{A_{\text{shaft}}}$
Bending	$\sigma_{MB} = \frac{\sqrt{M_{xB}^2 + M_{yB}^2}}{W_B}$	$\sigma_{MB} = \frac{M_{xB}}{W_{xB}} + \frac{M_{yB}}{W_{yB}}$	$\sigma_{\text{M-shaft}} = \frac{M_{\text{shaft}}}{W_{\text{shaft}}}$
Shear	Negligible	Negligible	$\tau_{\text{M-shaft}} = \frac{M_{\text{x-shaft}}}{2W_{\text{shaft}}}$
Combined (axial + bending)	$\sigma_{\text{eq}B} = \sigma_{zB} + \sigma_{MB}$		$\sigma_{\text{eq}} = \sqrt{(\sigma_{\text{x-shaft}} + \sigma_{\text{M-shaft}})^2 + 3\tau_{\text{M-shaft}}^2}$

Table 9.7 Partial safety factors for SLM loads from IEC 61400-2

Load determination method	Safety factor for fatigue loads, γ_f	Safety factor for ultimate loads, γ_f
Simple load calculation	1.0	3.0
Aero-elastic modelling with design data (rpm, power)	1.0	1.35
Load measurements with extrapolation	1.0	3.0

Table 9.8 Partial safety factors for material characterisation form IEC 61400-2

Condition	Full characterisation	Minimal characterisation
Fatigue strength	1.25	10.0
Ultimate strength	1.1	3.0

9.3.2 Partial Safety Factors

Ref. IEC 61400-2 defines partial safety factors for the load case calculations based upon the analysis method used. These factors, shown in Table 9.7, reflect the judgement that the SLM is less accurate than a full aero-elastic analysis. The second set of partial safety factors for the component materials are shown in Table 9.8. To use the full characterisation factors, i.e. the lower safety factors, the material properties must have been estimated with 95% probability at a 95% confidence level. The other requirements are:

- the properties have been obtained from materials and configurations representative of the final structure,
- the test samples come from the same process as the final product,
- static and fatigue testing have been done with appropriate load spectrum and rate effects,
- environmental effects, such as ultra-violet degradation, have been included, and
- any appropriate geometry effects, such as fibreglass orientation in composite blades, have been adequately accounted for.

9.3.3 Ultimate Stress Analysis

The safety factors and the ultimate material strength determine the allowable design stress, σ_d, in any component according to[31]:

[31] (IEC 47).

$$\sigma_d = \frac{f_k}{\gamma_m \gamma_f} \tag{9.30}$$

where f_k is the ultimate material strength of the component, taken here to be the yield stress, γ_f is the load case partial safety factor as determined in Table 9.7 and γ_m is the relevant partial safety factor for the material from Table 9.8. In other words, the two partial safety factors are multiplied to give the final safety factor. For the design of a particular component to be deemed safe, then the equivalent component stress level from Table 9.6 must be lower than the allowable material stress limit, i.e.:

$$\sigma_{\text{eq}} < \sigma_d \tag{9.31}$$

9.3.4 Fatigue Failure Analysis

Load case A in the SLM, requires assessing the fatigue damage. IEC 61400-2 states that this is to be done using Miner's rule. The damage calculation is as follows[32]:

$$\text{Damage} = \sum_i \frac{n_i}{N_{\text{cycles}}\left(\gamma_f \gamma_m s_i\right)} \leq 1 \tag{9.32}$$

where n_i is the number of fatigue cycles in bin i of the characteristic load spectrum, s_i is the stress level of the fatigue cycles including effects from both mean and cyclic stress levels, N_{cycles}, is the number of cycles to failure as a function of the stress, which in turn is calculated using the same combined safety factor $\gamma_m \gamma_f$ for the loads and materials as was used in (9.30). The term in parenthesis in the denominator is called the "associated stress level". For the SLM, only case A considers fatigue and there is only one "bin", whose number of fatigue cycles is to calculated as[33]

$$n = N\,\Omega_{\text{design}} T_d / 60 \tag{9.33}$$

and T_d is the design life of the turbine in seconds.

By Miner's rule, a component will fail if the damage over the component's lifetime reaches unity.

Finally, the standard requires the designer to undertake a "critical deflection analysis" to ensure that no component deflection under load will compromise

---

[32] (IEC 48).

[33] (IEC 49).

safety. The most obvious possibility here is that the deflection of the blades under load will cause them to hit the tower.

9.4 Simple Load Model Analysis of 500 W Turbine

This section describes the determination of the forces and moments on the components of the turbine shown in Fig. 6.1 and converts them into equivalent stresses. The basic operating parameters are:

$$\begin{aligned} P_{\text{design}} &= 500\,\text{W} \\ \Omega_{\text{design}} &= 700\,\text{rpm} = 73.30\,\text{rad/s} \text{ @ } U_{\text{design}} = 10.5\ \text{m/s} \\ \Omega_{\text{max}} &= 1000\,\text{rpm} = 104.72\,\text{rad/s}. \end{aligned} \tag{9.34}$$

The design power, P_{design}, design wind speed, U_{design}, and angular velocity, Ω_{design}, are obviously set by the blade design for power extraction. The second implies $U_{\text{ave}} = 7.5$ m/s, and from Table 9.2 of IEC 61400-2, $U_{\text{ref}} = 37.5$ m/s and $U_{e50} = 1.4U_{\text{ref}}$ by (IEC 10) $= 52.5$ m/s. The third is an estimate of the maximum blade speed from all the load cases.

λ_{design} can be calculated using Eq. 9.29 noting that the blade radius is actually 0.97 m:

$$\lambda_{\text{design}} = \frac{0.97}{10.5}\frac{\pi \times 700}{30} = 6.77 \tag{9.35}$$

From Eq. 9.2, noting that power must be in kW, the efficiency η is

$$\eta = 0.6 + 0.005 \times 0.5 = 0.603 \tag{9.36}$$

Now (9.1) can be used to find Q_{design}

$$Q_{\text{design}} = 30 \times 0.5 \times 1000/(0.603 \times \pi \times 700) = 11.32\,\text{Nm} \tag{9.37}$$

From (9.3) $\omega_{\text{yaw,max}}$ is given by:

$$\omega_{\text{max}} = 3 - 0.01(\pi \times 0.97^2 - 2) = 2.99\,\text{rad/s} \tag{9.38}$$

9.4.1 Loads for Case A: Normal Operation

The centrifugal load is given by Eq. 9.4 which involves two parameters that have not yet been determined. m_B, the blade mass, is easily found by weighing, or for pre-manufacturing analysis, from finite element or CAD software. Alternatively, the analysis in Sect. 6.4 for the moment of inertia can be easily modified to give m_B. For the current timber blades with density 550 kg/m^3, $m_B = 0.4$ kg. The blade centre of mass, R_{cog}, can be determined in similar fashion, taking note that the hub end of the blade is not likely to correspond to zero radius. In this case $R_{\text{cog}} = 0.379$ m. Thus

$$\Delta F_{zB} = 2 \times 0.4 \times 0.379 \times 73.30^2 = 1629.23\,\text{N} \tag{9.39}$$

From (9.5) and (9.6) respectively:

$$\Delta M_{xB} = 11.32/3 + 2 \times 0.4 \times 9.81 \times 0.379 = 6.75\,\text{Nm} \tag{9.40}$$

and

$$\Delta M_{yB} = 7.11 \times 11.32/3 = 25.55\,\text{Nm} \tag{9.41}$$

Equation 9.7 gives the shaft load as:

$$\Delta F_{\text{x-shaft}} = \frac{3 \times 7.11 \times 11.32}{2 \times 0.97} = 118.55\,\text{N} \tag{9.42}$$

From (9.8), using $e_r = 0.005 \times 0.97 = 0.0049$ m, and a measured rotor mass $m_r = 3.98$ kg,

$$\Delta M_{\text{x-shaft}} = 11.32 + 2 \times 3.98 \times 9.81 \times 0.0049 = 11.70\,\text{Nm} \tag{9.43}$$

so the assumed eccentricity has little effect in this case. The other shaft moment is given by (9.9) using the measured (or designed) length from the rotor to the first bearing, $L_{rb} = 0.026$ m,

$$\Delta M_{\text{shaft}} = 2 \times 3.98 \times 9.81 \times 0.026 + \frac{0.97}{6} \times 118.55 = 21.20\,\text{Nm} \tag{9.44}$$

9.4.2 *Loads for Case B: Yawing*

The blade root bending moment depends on the additional parameters $L_{rt} = 0.218$ m, the distance from the rotor to the yaw axis, and J_B, the blade moment of inertia. From the CAD model of the blade, $J_B = 0.0633$ kg m^2 and this value was confirmed by tests described by Wright [5].[34] Note also that J_B can be approximated as $J_B \approx m_B R_{\text{cog}}^2 = 0.4 \times 0.379^2 = 0.0575\,\text{kgm}^2$ for the present rotor. From (9.10):

[34] Note that this value of J for wooden blades is considerable less than for the composite blades of the same shape considered in Chap. 6.

$$M_{yB} = 0.4 \times 2.99^2 \times 0.218 \times 0.379 + 2 \times 2.99 \times 0.0633 \times 73.30 + \frac{0.97}{9} \times 118.55$$
$$= 40.83\,\text{Nm} \tag{9.45}$$

Similarly, (9.12) gives the shaft moment as

$$M_{\text{shaft}} = 3 \times 2.99 \times 73.30 \times 0.0633 + 3.98 \times 9.81 \times 0.0260 + \frac{0.97}{6} \times 118.55$$
$$= 61.81\,\text{Nm} \tag{9.46}$$

In both (9.45) and (9.46), the gyroscopic terms are responsible for over two-thirds of the moment.

9.4.3 Loads for Case C: Yaw Error

Eq. 9.13 for this load requires avalue for $A_{\text{proj},B}$. This can be done using a modification of the analysis in Sect. 6.4. For the three blades, $A_{\text{proj},B} = 0.231\ \text{m}^2$ for zero yaw. With $C_{l,\max} = 2.0$, Eq. 9.13 gives

$$M_{y,B} = \frac{1}{8} \times 1.2250 \times 0.231 \times 2 \times 0.97^3 \times 73.30^2 \left[1 + \frac{4}{3 \times 6.77} + \left(\frac{1}{6.77}\right)^2\right]$$
$$= 422.82\,\text{Nm} \tag{9.47}$$

9.4.4 Loads for Case D: Maximum Thrust

Using Eq. 9.14 the calculation is straightforward:

$$F_{x-\text{shaft}} = 3.125 \times 0.5 \times 1.225 \times 7.5^2\pi \times 0.97^2 = 318.25\,\text{N} \tag{9.48}$$

9.4.5 Loads for Case E: Maximum Rotational Speed

The calculation of the centrifugal loading at the blade root is also straightforward. From (9.15)

$$F_{zB} = 0.4 \times 104.72^2 \times 0.379 = 1662.48\,\text{N} \tag{9.49}$$

The shaft bending moment is found from Eq. 9.16:

$$\begin{aligned} M_{\text{shaft}} &= 3.98 \times 9.81 \times 0.026 + 3.98 \times 0.0049 \times 104.72^2 \times 0.026 \\ &= 6.52\,\text{Nm} \end{aligned} \tag{9.50}$$

9.4.6 Loads for Case F: Short at Electrical Connection

Now consider the load due to a short circuit at the generator. The turbine has a permanent generator, so $G = 2$ (the short circuit torque factor). Thus,

$$M_{\text{x-shaft}} = 2 \times 11.32 = 22.64\,\text{Nm} \tag{9.51}$$

from (9.17) and

$$M_{xB} = 22.64/3 = 7.55\,\text{Nm} \tag{9.52}$$

from (9.18).

Load case G does not apply to this turbine, as it does not use any form of braking.

9.4.7 Loads for Case H: Parked Wind Loading

This wind loading is applied to the parked turbine at a wind speed of U_{e50}. It will be assumed that the turbine has furled and the blades are effectively stationary. Because of the furling the blades will not be normal to the wind direction, so if we use (9.23) with a projected blade area of 0.231 m^2, then the resulting bending moment will be over-estimated:

$$M_{yB} = \frac{1}{4} \times 1.5 \times 1.225 \times 52.5^2 \times 0.231 \times 0.97 = 283.71\,\text{Nm} \tag{9.53}$$

Similarly, from (9.25)

$$F_{\text{x-shaft}} = 0.5 \times 3 \times 1.5 \times 1.225 \times 52.5^2 \times 0.231 = 1754.89\,\text{N} \tag{9.54}$$

where C_d is taken as 1.5 from Table 9.5.

9.5 Equivalent Component Stresses and Ultimate Material Strengths

The loads that have been calculated in Sect. 9.4 must be converted to equivalent stresses. This requires additional information, principally of component areas with the generic symbol A, and values for the section modulus, W, the second moment

divided by the distance from the centroid to the point of maximum stress. This is the furthest point on the component from the neutral axis, which runs through the centre of mass. Use can then be made of the formulae in Table 9.6 to evaluate the stresses. For example, r_{shaft}, the radius of the rotor shaft is 12.5 mm, so $A_{\text{shaft}} = 4.909 \times 10^{-4}$ m^2. The second moment of inertia for the circular shaft is $\pi r_{\text{shaft}}^4/4 = 1.918 \times 10^{-8}\,\text{m}^4$, and $W_{\text{shaft}} = 1.9175 \times 10^{-8}/0.0125 = 1.534 \times 10^{-6}$ m^3.

The corresponding information for the blade is harder to obtain. The blade root area, A_B, was taken as the smallest cross-sectional area in the root region as determined from a CAD model of the blade. This gave $A_B = 0.00146$ m^2. This occurred just above the 0.095 m (in the chord direction) by 0.105 m (radial) by 0.025 m thick rectangular attachment section.

The CAD software was used to determine the other moments about the axes defined in Appendix 2 of IEC 61400-2:

$$I_{xx} = 2.79 \times 10^{-8}\,\text{m}^4 \quad \text{and} \quad I_{yy} = 1.199 \times 10^{-6}\,\text{m}^4 \tag{9.55}$$

and the respective distances from the stress points were found to be

$$c_{xB} = 0.00919\,\text{m} \quad \text{and} \quad c_{yB} = 0.00778\,\text{m} \tag{9.56}$$

Thus $W_{xB} = 2.79 \times 10^{-8}/0.00919 = 3.0359 \times 10^{-6}$ m^3 and $W_{yB} = 1.19 \times 10^{-6}/0.00778 = 1.527 \times 10^{-4}$ m^3.

The ultimate material strength, denoted f_k in (9.30), of the timber blades is 123 MPa, Peterson and Clausen [6]. The generator shaft is assumed to be made from mild steel for which $f_k = 250$ MPa.

The next subsections show the calculations of the equivalent stresses. It is assumed that the shaft and other metal components are made from structural steel with a yield stress of 350 MPa and the timber blades have a yield stress of 120 MPa [6]. Both materials are assumed to be well characterised so that the appropriate material safety factors are 1.25 for fatigue and 1.1 for ultimate strength. Using the load safety factor of 3 for the ultimate strength, the blades will be safe in ultimate loads if the equivalent stress is less than 36.4 MPa, and the steel components if the stress is less than 106.1 MPa.

9.5.1 Equivalent Stress for Case A: Normal Operation

Load case A is the only fatigue load. However, it will be shown below that the equivalent average stress level is still required to complete the analysis. The equivalent stress in the blade root is found using Eqs. 9.39–9.41 inclusive. Since the root area of the blade in Fig. 6.1 is more rectangular than circular, the equations in the second column of Table 9.6 will be used. Thus

$$\begin{aligned}\sigma_{\mathrm{eq}} &= \frac{\Delta F_{zB}}{A_B} + \frac{\Delta M_{xB}}{W_{xB}} + \frac{\Delta M_{yB}}{W_{yB}} \\ &= \frac{1629.23}{1.46 \times 10^{-3}} + \frac{6.75}{3.036 \times 10^{-6}} + \frac{25.55}{1.527 \times 10^{-5}} \\ &= 3.51\ \mathrm{MPa}\end{aligned} \tag{9.57}$$

For the main shaft, Eqs. 9.42–9.44 and the equations in the last column of Table 9.6 give:

$$\begin{aligned}\sigma_{\mathrm{eq}} &= \sqrt{\left(\frac{\Delta F_{\mathrm{x-shaft}}}{A_{\mathrm{shaft}}} + \frac{\Delta M_{\mathrm{shaft}}}{W_{\mathrm{shaft}}}\right)^2 + 3\left(\frac{\Delta M_{\mathrm{x-shaft}}}{2W_b}\right)^2} \\ &= \sqrt{\left(\frac{118.55}{4.909 \times 10^{-4}} + \frac{21.20}{1.534 \times 10^{-6}}\right)^2 + 3\left(\frac{11.70}{2 \times 1.534 \times 10^{-6}}\right)^2} \\ &= 15.53\ \mathrm{MPa}\end{aligned} \tag{9.58}$$

In Eqs. 9.57 and 9.58, the equivalent stress was calculated from the peak-to-peak variation in the fatigue load. However, *S,N*-diagrams for fatigue show the *amplitude* of the maximum stress versus the number of cycles for use in Miner's Rule. Thus the results from (9.57) and (9.58) must be divided by 2. Since $0.5 \times 15.53 = 7.77$ MPa is well below the endurance limit for steel, the shaft seems safe in fatigue. But there is a further complication: Miner's Rule, Eq. 9.32, assumes no average stress, whereas there will be mean equivalent stresses for Load Case A and this reduces fatigue strength. Even though IEC 61400-2 states that "mean values of the load ranges can be ignored" it is advisable to check the effect of the mean stress. This equals the equivalent stress for the peak-to-peak loads for the turbine cycling between 50 and 150% of the design speed. Then it is possible to use Gerber's relation, see, for example, Eq. 6.4 of Schijve [7] to estimate the maximum stress $\sigma_{\mathrm{eq,max}}$ for a particular number of cycles which will be less than the value if there were no mean stress. Assuming n is large enough for the endurance limit stress, σ_l, to be applicable then the equation for $\sigma_{\mathrm{eq,max}}$ which must replace σ_l, is

$$\frac{\sigma_{\mathrm{eq,max}}}{\sigma_l} + \frac{\sigma^2_{\mathrm{eq,max}}}{F_y^2} = 1 \tag{9.59a}$$

or

$$\sigma_{\mathrm{eq,max}} = \frac{F_y}{2}\left(\sqrt{\left(\frac{F_y}{\sigma_l}\right)^2 + 4} - \frac{F_y}{\sigma_l}\right) \tag{9.59b}$$

where F_y is the yield stress equal to 350 MPa in this case. If $\sigma_l = 200$ MPa, then (9.59b) gives $\sigma_{\mathrm{eq,max}} = 158$ MPa. Using the material load factor of 1.25, the maximum allowable stress is 158/1.25 = 127 MPa, so the shaft is easily safe.

9.5.2 Equivalent Stress for Case B: Yawing

These calculations are straightforward. For the blade root:

$$\sigma_{eq} = \frac{M_{yB}}{W_{yB}} = \frac{40.83}{1.527 \times 10^{-4}} = 0.27\,\text{MPa} \tag{9.60}$$

from (9.45). Equation 9.46 gives the maximum stress in the shaft as

$$\sigma_{eq} = \frac{M_{\text{shaft}}}{W_{\text{shaft}}} = \frac{61.81}{1.534 \times 10^{-6}} = 40.29\,\text{MPa} \tag{9.61}$$

9.5.3 Equivalent Stress for Case C: Yaw Error

Load case C considers only the bending moment on the blades, which is, using (9.47);

$$\sigma_{eq} = \frac{M_y}{W_{yB}} = \frac{422.82}{1.527 \times 10^{-4}} = 2.77\,\text{MPa} \tag{9.62}$$

9.5.4 Equivalent Stress for Case D: Maximum Thrust

As for load case C, D only considers a load on the main shaft, with the equivalent stress level calculated from (9.48);

$$\sigma_{eq} = \frac{F_{\text{x-shaft}}}{A_{\text{shaft}}} = \frac{318.25}{4.909 \times 10^{-4}} = 0.65\,\text{MPa} \tag{9.63}$$

9.5.5 Equivalent Stress for Case E: Maximum Rotational Speed

The loading for this case again considers only one load for both the main shaft and the blade root. The equivalent stress for the blade root, using the result of Eq. 9.49, is

$$\sigma_{eq} = \frac{F_{zB}}{A_B} = \frac{1662.48}{1.46 \times 10^{-3}} = 1.14\,\text{MPa} \tag{9.64}$$

Using the calculated moment from Eq. 9.50 the equivalent stress in the main shaft is;

$$\sigma_{eq} = \frac{M_{shaft}}{W_{shaft}} = \frac{6.52}{1.534 \times 10^{-6}} = 4.25\,\text{MPa} \tag{9.65}$$

9.5.6 Equivalent Stress for Case F: Short at Electrical Connection

Load case F also considers a single load for both the main shaft and the blade root. Again, the calculations for the equivalent stress are straightforward. For the blade root:

$$\sigma_{eq} = \frac{M_{xB}}{W_{xB}} = \frac{7.55}{3.036 \times 10^{-6}} = 2.49\,\text{MPa} \tag{9.66}$$

using the result of Eq. 9.52. For the main shaft:

$$\sigma_{eq} = \sqrt{3}\frac{M_{shaft}}{2W_{shaft}} = \frac{\sqrt{3} \times 22.64}{2 \times 1.534 \times 10^{-6}} = 12.78\,\text{MPa} \tag{9.67}$$

from (9.51).

9.5.7 Equivalent Stress for Load Case H: Parked Wind Loading

This case also considers one load on both the blade root and the main shaft. The equivalent stress levels for the blade root are

$$\sigma_{eq} = \frac{M_{yB}}{W_{yB}} = \frac{283.71}{1.527 \times 10^{-4}} = 1.86\,\text{MPa} \tag{9.68}$$

using (9.53). Similarly, (9.54) leads to:

$$\sigma_{eq} = \frac{F_{x-shaft}}{A_{shaft}} = \frac{1754.89}{4.909 \times 10^{-4}} = 3.58\,\text{MPa} \tag{9.69}$$

for the main rotor shaft.

9.6 Spreadsheet for the Simple Load Model

To simplify the use of the SLM, an excel spreadsheet, SLM_500.xls has been developed. Figure 9.1 shows the first chart in the spreadsheet. The shaded cells indicate values that must be input by the user in order to undertake the calculations

Spreadsheet for the Simple Load Model (SLM)			
Australian Windpower 500 W turbine			
Enter Input Data into Grey Cells Only			
Description	**Input Value**	**Units**	**Symbol**
Air density	1.2250	kg/m^3	ρ
Gravitational acceleration	9.8100	m/s^2	g
Reference Wind Speed	37.5000	m/s	V_ref
Average Wind Speed	7.5000	m/s	V_ave
Number of Blades	3	n/a	B
Blade Tip Radius	0.9700	m	R
Total Planform Area of the Blade	0.2310	m^2	A_proj,B
Drag Coefficient of the Blades	1.5000	n/a	C_d
Max Lift Coefficient of the Blades	2.0000	n/a	C_l,max
Thrust Coefficient	0.5000	n/a	C_T
Maximum Rotor Speed	1000.0000	rpm	n_max
Design Rotor Speed	700.0000	rpm	n_design
Second Moment of Inertia for each Blade	0.0633	kgm^2	I_B
Single Blade Mass	0.4000	kg	m_B
Rotor Mass (All Blades plus Hub)	3.9800	kg	m_r
Distance from Blade centre of gravity to rotor axis	0.3790	m	R_cog
Distance Between Rotor Centre and First Bearing	0.0260	m	L_rb
Distance Between the Rotor Centre and the yaw axis	0.2180	m	L_rt
Gearbox Ratio (enter 1.0 for no gearbox)	1.0000	n/a	Gear
Enter "Y" if brake is on high speed side of gearbox, otherwise "N"	N	n/a	n/a
Brake torque (enter 0.0 for no brake)	0.0000	Nm	M_brake
Design Power	0.5000	kW	P_design
Short Circuit Torque Factor	2.0000	n/a	G
Type "Y" if blades are stationary during parking, otherwise "N"	Y	n/a	n/a

Fig. 9.1 The first sheet of the SLM spreadsheet

as described in the previous section. The values in the white cells are prescribed by IEC 61400-2. The second chart, Fig. 9.2, displays the parameters calculated from the input information (Fig. 9.3).

The third chart labeled "Loads from SLM" contains the SLM loads with reference to the relevant equation in IEC 61400-2 1. The fourth chart, Fig. 9.4, allows the user to enter the extra information required to convert the SLM loads into equivalent stresses. The fatigue data was taken from Peterson and Clausen [6] for Hoop pine, a native Australian timber that grows well in plantation.

Parameters Calculated from Input Data			
Description	**Value**	**Units**	**Symbol**
Design Wind Speed	10.50	m/s	V_design
50 year extreme wind speed	52.50	m/s	V_e50
50yr extreme tip speed ratio	1.93	n/a	λ_e50
Design Tip Speed Ratio	6.77	n/a	λ_design
Drive Train Efficiency	0.60	n/a	'H
Design Torque	11.32	Nm	Q_design
Projected Area (turbine swept area)	2.96	m^2	A_proj
Design Rotational Speed of the Rotor	73.30	rad/s	ω_n,design
Maximum Possible Rotor Speed	104.72	rad/s	ω_n,max
Max Yaw Rate	2.99	rad/s	ω_yaw,max
50yr extreme tip speed ratio	1.93	n/a	λ_e50
Eccentricity of the Rotor Centre of Mass	4.85E-03	m	e_r
Effective brake torque	0.00	Nm	M_brake

Fig. 9.2 The second sheet of the SLM spreadsheet

The fifth chart in Fig. 9.5 gives the output of equivalent stresses. Figure 9.6 shows the sixth chart used to determine the material strengths for ultimate loading and fatigue using the input information on material characterisation and number of cycles to fatigue. It has been assumed that the materials for the blade and the main shaft have not been properly characterised and so attract the highest partial load factors from Tables 9.7 and 9.8. The sixth chart completes the data entry.

The final chart in Fig. 9.7 shows the outcome of the SLM calculations in terms of the safety of the relevant component (blade root or rotor shaft). The rotor shaft fails load case B, whereas the shaft and the blades easily pass all other load cases.

A major part of the reason why the rotor shaft fails Case B is the inadequate characterisation of its properties. The assumed yield stress of 250 MPa is much lower than is usual for case-hardened steels often used for shafts. In any case, even for this low value, being able to demonstrate proper characterisation would lower the material safety factor from 3 to 1.1 and this alone would allow the shaft to pass.

9.7 Further Test Requirements

IEC 61400-2 mandates testing to determine or validate the design power, design rotational speed, design shaft torque, and maximum rotational speed. In addition the following are required:

- a blade static test, combining the highest centrifugal load with the largest bending moment to test the integrity of the blade. The standard recommends fatigue testing of the blade, but does not require it.

Loads from SLM				
Load Case A - Fatigue Loads on Blades and Rotor Shaft				
Equation	**Description**	**SLM Value**	**Units**	**Symbol**
	Blade Loads			
9.4 (IEC 21)	Centrifugal Force at the Blade Root (z-axis)	1629.23	N	ΔF_zB
9.5 (IEC 22)	Lead-lag Root Bending Moment (x-axis)	6.75	Nm	ΔM_xB
9.6 (IEC 23)	Flapwise Root Bending Moment (y-axis)	25.55	Nm	ΔM_yB
	Shaft Loads			
9.7 (IEC 24)	Thrust on shaft (x-axis)	118.55	N	ΔF_x-shaft
9.8 (IEC 25)	Shaft Moment about x-axis	11.70	Nm	ΔM_x-shaft
9.9 (IEC 26)	Shaft Moment	21.20	Nm	ΔM_shaft
Load Case B - Blade and Rotor Shaft Loads during Yaw				
Equation	**Description**	**SLM Value**	**Units**	**Symbol**
9.16 (IEC 28)	Flapwise Root Bending Moment (y-axis)	40.83	Nm	M_yB
9.11,12 (IEC 29,30)	Bending moment on the shaft	61.81	Nm	M_shaft
Load Case C - Yaw Error Load on Blades				
Equation	**Description**	**SLM Value**	**Units**	**Symbol**
9.13 (IEC 31)	Flapwise Root Bending Moment (y-axis)	422.82	Nm	M_yB
Load Case D - Maximum Thrust on Shaft				
Equation	**Description**	**SLM Value**	**Units**	**Symbol**
9.14 (IEC 32)	Maximum Thrust on Shaft	318.25	N	F_x-shaft
Load Case E - Maximum Rotational Speed				
Equation	**Description**	**SLM Value**	**Units**	**Symbol**
9.15 (IEC 33)	Centrifugal Force at the Blade Root (z-axis)	1662.48	N	F_zB
9.16 (IEC 34)	Bending Moment on the shaft	6.52	Nm	M_shaft
Load Case F - Short at Load Connection				
Equation	**Description**	**SLM Value**	**Units**	**Symbol**
9.17 (IEC 35)	Bending Moment on Shaft	22.64	Nm	M_x-shaft
9.18 (IEC 36)	Lead-lag Root Bending Moment (x-axis)	7.55	Nm	M_xB
Load Case G - Shutdown Braking				
Equation	**Description**	**SLM Value**	**Units**	**Symbol**
9.19 (IEC 37)	Bending Moment on Shaft	n/a	Nm	M_x-shaft
9.20 (IEC 38)	Lead-lag Root Bending Moment (x-axis)	n/a	Nm	M_xB
Load Case H - Parked Wind Loads during Idling				
Equation	**Description**	**SLM Value**	**Units**	**Symbol**
9.23,24 (IEC 39,40)	Flapwise Root Bending Moment (y-axis)	283.71	Nm	M_yB
9.25,26 (IEC 41,42)	Maximum Thrust on Shaft	1754.89	N	F_x-shaft

Fig. 9.3 The third sheet of the SLM spreadsheet

Enter Additional Data into Grey Cells Only			
Description	**Value**	**Units**	**Symbol**
Diameter of the Shaft	2.5000E-02	m	
Cross Sectional Area of the Shaft	4.9087E-04	m^2	A_shaft
The second moment of inertia for the shaft	1.9175E-08	m^4	I_x-shaft
Section modulus for the shaft	1.5340E-06	m^3	W_shaft
Cross Sectional Area of the Blade Root	1.4600E-03	m^2	A_B
Ixx for the blade	2.7900E-08	m^4	I_xxB
x-distance from blade centroid to the maximum stress point	9.7100E-03	m	c_xB
Iyy for the blade	1.1990E-06	m^4	I_yyB
y-distance from blade centroid to the maximum stress point	7.7800E-03	m	c_yB
Blade x-section modulus	3.0359E-06	m^3	W_xB
Blade y-section modulus	1.5270E-04	m^3	W_yB
Ultimate Material Strength for the Blades	1.2000E+02	MPa	f_kB
Ultimate Material Strength for the Shaft	2.5000E+02	MPa	f_k-shaft
Enter Additional Data for Fatigue Calculations into Yellow Cells			
Description	**Value**	**Units**	**Symbol**
Design life of the turbine	20	Years	
	631152000	s	T_d
Number of Fatigue Cycles	2.21E+10	n/a	n_i
Number of Cycles to Failure as a Function of Stress (Shaft)	1.00E+10	n/a	N_shaft
Number of Cycles to Failure as a Function of Stress (Blade)	1.23E+13	n/a	N_blade

Fig. 9.4 The fourth sheet of the SLM spreadsheet

- a duration test to investigate the structural integrity and material degradation, the quality of the environmental protection, and the turbine's dynamic behaviour. The duration test must take at least 6 months and consist of over 100 days of accumulated operation. Field testing of the 500 W turbine to determine the instantaneous blade loads and their analysis is described in Wilson and Clausen [8, 9].
- a safety and function test to investigate the control and protection system. This might include testing the overspeed protection, emergency shutdown etc.

The standard allows the turbine's maximum yaw rate to be measured but Equation (IEC 27) must always be used in the SLM.

9.8 Final Remarks

It is unlikely that many manufacturers of small turbines will go to the effort and expense of certification to the IEC standard. Nevertheless, IEC 61400-2 provides a very useful methodology for designing safe small turbines and the spreadsheet

Fig. 9.5 The fifth sheet of the SLM spreadsheet

Calculation of the Equivalent Stresses	
Description	**Equivalent Stress (MPa)**
Load Case A - Fatigue Loads on Blades and Rotor Shaft	
Blades	3.51
Shaft	15.53
Load Case B - Blade and Rotor Shaft Loads during Yaw	
Blades	0.27
Shaft	40.29
Load Case C - Yaw Error Load on Blades	
Blades	2.77
Load Case D - Maximum Thrust on Shaft	
Shaft	0.65
Load Case E - Maximum Rotational Speed	
Blades	1.14
Shaft	4.25
Load Case F - Short at Load Connection	
Blades	2.49
Shaft	12.78
Load Case G - Shutdown Braking	
Blades	n/a
Shaft	n/a
Load Case H - Parked Wind Loads during Idling	
Blades	1.86
Shaft	3.58

described in this paper facilitates the analysis. Addressing the limitations of the standard will greatly add to our knowledge of small wind turbine behaviour.

It should be apparent from the previous sections that designing small turbines to IEC 61400-2, and then demonstrating that the turbine is safe is not simple. The SLM calculations of the 500 W turbine highlight the critical role of the turbine's yaw behaviour. The largest load was due to the Coriolis acceleration on the rotating and yawing blades.

Part of the reason for the single failure was the poor state of knowledge of the shaft material and it would be more than worthwhile to improve this situation. It is also possible that the standard over-estimates the Coriolis component of the shaft load because of the mandatory formulation of the maximum yaw rate and the requirement that this be combined with the design rotor speed.

Partial Safety Factors (PSF) for SLM			
Description	**PSF**	**Symbol**	
Partial Safety Factor for Fatigue Loads	1.00	g_f-f	
Partial Safety Factor for Ultimate Loading	3.00	g_f-u	
Calculations for the material PSF (Enter Data into Grey Cells)			
Description	**Input**	**PSF**	**Symbol**
Enter "Y" for full characterisation of blade material, otherwise "N"	N		
Blade Fatigue strength partial safety factor	-->	10	γ_mB-f
Blade Ultimate strength partial safety factor	-->	3	γ_μB–υ
Enter "Y" for full characterisation of shaft material, otherwise "N"	N		
Shaft Fatigue strength partial safety factor	-->	10	γ_m-shaft-f
Shaft Ultimate strength partial safety factor	-->	3	γ_m-shaft-u
Material Strengths (with Factors of Safety)			
Description	**Value**	**Units**	
Ultimate Strength for Blade Material	13.33	MPa	
Ultimate Strength for Shaft Material	27.78	MPa	
Calculations for Fatigue (Enter Data into Grey Cells)			
Description	**Value**	**Units**	**Symbol**
Blade Stress Level	3.51	MPa	s_iB
Associated Blade Stress Level	35.06	MPa	
Number of Cycles to Failure at this Stress	9.81E+15	-	N_B
Shaft Stress Level	15.53	MPa	s_i-shaft
Associated Shaft Stress Level	155.34	MPa	-
Number of Cycles to Failure at this Stress	inf	-	N_shaft

Fig. 9.6 The sixth sheet of the SLM spreadsheet

The SLM offers a much easier framework for small wind turbine design than the expensive alternative of aero-elastic modeling that is routinely pursued for large machines. It is in everyone's interests to continue to provide data for, and testing of, the simple load model to improve its applicability while retaining its simplicity.

Another necessary development of the standard is to address the partial load factors in Table 9.7. The high values used in the SLM reflect the simplistic nature of the equations used. However, some terms in those equations, such as the Coriolis terms in (IEC 28–30) are based on exact equations and so should attract lower safety factors when our knowledge of yaw behaviour has improved sufficiently to make firmer judgements on the maximum yaw rate.

9.8.1 Further Reading

The SLM is a good starting point for turbine structural and dynamic analysis, but it is only a starting point. Aero-elastic modeling has been used for small turbines and

Simple Load Model Results			
Load Case A - Fatigue Loads on Blades and Rotor Shaft			
	Fatigue Damage Limit	**Fatigue Damage**	**Conclusion**
Blades	1.00	2.25E-06	SAFE
Shaft	1.00	Infinite Life	SAFE
Load Case B - Blade and Rotor Shaft Loads during Yaw			
	Material Stress Limit (MPa)	**Calculated Stress (MPa)**	**Conclusion**
Blades	13.33	0.27	SAFE
Shaft	27.78	40.29	FAIL
Load Case C - Yaw Error Load on Blades			
	Material Stress Limit (MPa)	**Calculated Stress (MPa)**	**Conclusion**
Blades	13.33	2.77	SAFE
Load Case D - Maximum Thrust on Shaft			
	Material Stress Limit (MPa)	**Calculated Stress (MPa)**	**Conclusion**
Shaft	13.33	0.65	SAFE
Load Case E - Maximum Rotational Speed			
	Material Stress Limit (MPa)	**Calculated Stress (MPa)**	**Conclusion**
Blades	13.33	4.25	SAFE
Shaft	27.78	1.14	SAFE
Load Case F - Short at Load Connection			
	Material Stress Limit (MPa)	**Calculated Stress (MPa)**	**Conclusion**
Blades	13.33	2.49	SAFE
Shaft	27.78	12.78	SAFE
Load Case G - Shutdown Braking			
	Material Stress Limit (MPa)	**Calculated Stress (MPa)**	**Conclusion**
Blades	13.33	n/a	n/a
Shaft	27.78	n/a	n/a
Load Case H - Parked Wind Loads during Idling			
	Material Stress Limit (MPa)	**Calculated Stress (MPa)**	**Conclusion**
Blades	13.33	3.58	SAFE
Shaft	27.78	1.86	SAFE

Fig. 9.7 The seventh sheet of the SLM spreadsheet

will no doubt become more popular, but it is not easy to implement. Readers interested in pursuing this topic should start with the programs available from the excellent NREL site: http://wind.nrel.gov/designcodes/simulators/. Some of the impressive results that can be obtained from aero-elastic simulations are described in Jonkman and Hansen [10] and the NREL aero-elastic codes are described by Buhl and Manjock [11].

References

1. GL (2010) Guidelines for the certification of wind turbines, Germanischer Lloyd. https://www.gl-group.com/wind_guidelines/wind_guidelines.php?lang=en. (accessed 10 Oct 2010)
2. Burton T, Sharpe D, Jenkins N, Bossanyi E (2001) Wind energy handbook. Wiley, New York
3. Wood DH (2009) Using the IEC simple load model for small wind turbines. Wind Eng 33:139–154
4. Eggleston DM, Stoddard FS (1987) Wind turbine engineering design. van Nostrand, New York
5. Wright AK (2005) Aspects of the aerodynamics and operation of a small horizontal axis wind turbine. PhD thesis, School of Engineering, University of Newcastle
6. Peterson P, Clausen PD (2004) Timber for high efficiency small wind turbine blades. Wind Eng 28:87–96
7. Schijve J (2009) Fatigue of structures and materials. Springer, Dordrecht
8. Wilson SVR, Clausen PD (2007) Aspects of the dynamic response of a small wind turbine blade in highly turbulent flow: part 1 measured blade response. Wind Eng 31:1–16
9. Wilson SVR, Clausen PD (2007) Aspects of the dynamic response of a small wind turbine blade in highly turbulent flow: part 2 predicted blade response. Wind Eng 31:217–231
10. Jonkman JM, Hansen AC (2004) Development and validation of an aeroelastic model of a small furling wind turbine. NREL/CP-500-39589. http://www.nrel.gov/docs/fy05osti/39589.pdf. (accessed 17 May 2010)
11. Buhl ML, Manjock A (2006) A comparison of wind turbine aeroelastic codes used for certification, NREL/CP-500-39113. http://www.nrel.gov/wind/pdfs/39113.pdf. (accessed 14 Sept 2010)

Chapter 10
Tower Design and Manufacture

10.1 Introduction

As with other basic design decisions, such as the number of blades and whether the rotor is upwind or downwind, there is a larger variety of tower types for small turbines than for larger ones. This chapter concentrates on the main types: stand-alone monopoles, guyed poles or lattices, and non-guyed lattice towers of two main forms. Table 10.1 lists the main tower types with their advantages and disadvantages. Traditionally, lattice and guyed-towers have been the most common but the widespread deployment of grid-connected small turbines has lead to monopoles taking over, largely because they do not need guy-wires. Lattice towers come in two basic types: those made from steel angle or other sections which are assembled on site, and tubular lattice towers that are pre-fabricated in sections. Site issues, foundations, and tower loads during raising and lowering are covered in Chap. 12. This chapter concentrates on design and fabrication. The key design requirements are to ensure that the maximum design stress is below the allowable material stress, no tower member in compression will buckle, and the tower's lowest natural frequency is not likely to be significantly excited by blade passing frequencies. Note that the IEC Simple Load Model (SLM) described in Chap. 9 mandates only the first requirement. Some manufacturing issues such as corrosion resistance by galvanising will be described briefly.

Not all small wind turbines have towers, and some have shortened, or specially built ones: examples include building-mounted turbines and micro-turbines used for yachts. This feature of micro turbines is recognized in IEC 61400-2 which allows turbines of swept area less that 2 m^2 to be certified independently of the tower. Large wind turbine tower design is more complex than the simple use of safety factors described in this chapter, see, for example, Lavassas et al. [1] and Sect. 3 of Baniotopoulos [2].

The following IEC load cases are considered: Load Case A for fatigue during normal operation, B for loads during yaw, and Load Case H: Parked wind loading where it will be assumed that the rotor is stationary.

D. Wood, *Small Wind Turbines*, Green Energy and Technology,
DOI: 10.1007/978-1-84996-175-2_10, © Springer-Verlag London Limited 2011

The next section describes the design of monopole towers. They are the only type whose generic form can be analysed simply without finite element analysis (FEA). Subsequent sections will cover the remaining tower types, but in less depth.

For simplicity, it will be assumed that tower design starts after the main loads from the turbine-top—the platform, tail fin, blades etc.—are known. These can be calculated using the SLM as was done for other turbine components in Chap. 9. IEC 61400-2 states that "support structures shall also meet local codes and regulations". Many standards set the extreme wind speed as well as the methodology of determining the wind loads due to the drag on the tower and turbine. A summary of extreme wind values around the world is given in Appendix D of Holmes [3]. The value of 50 m/s used for the examples in this chapter is a typical one from the IEC safety standards for both large and small turbines. For example, a Class III wind turbine has a reference wind speed of 37.5 m/s. This value must be multiplied by 1.4 to get the 3-s gust "50 year return" wind speed which is 52.5 m/s from Table 9.2. In general, and obviously, wind turbines should be designed for the windiest possible location in the region covered by a particular local code. In some standards, the tower natural frequency determines the drag coefficient, and so influences the wind loads in a manner that has no counterpart in IEC 61400-2. The requirement to "meet local codes" makes it difficult to provide design guidelines that are universally valid. Fortunately, however, the methodology of many codes is similar and, in many cases, so is their outcome, Carril et al. [4] and Chap. 15 of Holmes [3]. A brief discussion of Eurocode 3 [5] for wind loading is given by Geurts and van Bentum [6] and its application to lattice towers is covered by Baniotopoulos [2].

The main tower-top loads are the maximum thrust, $T_{\max}$, and weight, m_{tt}. In addition, it is assumed that the tower height, h, has been decided. The tantalizing question of what is the best height for a particular installation is considered in Chap. 12 as are the additional loads during raising and lowering. These can be larger than the loads set by the extreme wind speed, even when raising and lowering occur only in calm weather. It is further assumed that the turbine connection to its tower fixes the minimum tower diameter, d_0, for any type.

For simplicity, the analysis will be restricted to static-linear behaviour. Thus the tower material is assumed to respond linearly to the imposed load and that the deflections are small enough to make the original (unloaded) shape sufficiently accurate for stress analysis. "In the analysis of towers [in general] the largest uncertainty is an accurate knowledge of the wind loads. Highly sophisticated methods of analysis cannot improve this. A static-linear-three dimensional structural analysis is sufficient for almost all lattice tower structures" [7]. "Three dimensional" means that the analysis must capture the actual tower shape. This is relatively easy to do with sufficiently powerful FEA software. In this chapter it will be assumed that all structural elements of a tower have the same material properties.

Designing for extreme or "ultimate" loads for which the drag coefficients are well known usually means that a lower safety factor can be used than those encountered for the IEC SLM in Chap. 9. In fact, all the wind loading standards known to the author stipulate safety factors lower than the SLM. Finally, it is noted that wind loads may be only one of a number of loads that must be considered in designing a tower. Additional loads may include those due to snow, ice, and earthquakes. These are not covered in this book but the basic methodology can be found in the standards referred to.

10.2 Monopole Towers

This analysis considers only the general form of the tower and no details such as the turbine mounting flange and baseplate. Other important details include the selection of the foundation bolts, the inclusion of an access panel, and possible provision of rib-stiffeners as used on large towers, e.g. Lavassas et al. [1] and Uys et al. [8]. A full three-dimensional FEA is usually required to design these components. Following the discussion at the end of the last section, the remaining parameters of the basic monopole design to be determined are the cross-sectional shape, the thickness, t, of the steel in each section, and the base diameter, d_h. These have to be chosen to withstand the loads from the turbine as well as the tower self-weight and the wind load. In addition it is necessary to calculate the horizontal force and base overturning moment before designing the foundation. The simple analysis that is now developed is used in Sect. 10.3 with an optimization process to determine the minimum mass tower which is probably the cheapest to make and transport. The analysis closely follows that of Kocer and Arora [9], Dicleli [10] and Clifton-Smith and Wood [11].

The horizontal force on the tower due to an extreme wind, with a typical speed of 50 m/s, is usually much larger than $T_{\max}$. This means that some care is required in selecting the tower shape to reduce drag. For example, a circular cross-section usually has a lower drag coefficient than, say, an octagonal section, see Table 9.5. On the other hand, tapered octagonal towers, similar to light poles, are easy to make. Figure 10.1 shows the bending of a half-section for a polygonal tower. The two halves are then seam welded to make a tapered section which should be hot-dipped galvanized. The sections can be slip-fitted together on site. This involves using a hand or hydraulic winch to force an overlap, typically of length 1.5 times the local diameter. For polygonal towers, the "diameter" will be taken to be the distance "across the flats" on the outside of the tower. For lighting towers, the slip fit sections rely on friction to remain connected but they should be bolted for wind turbine towers, to minimise movement due to fluctuating turbine loads and to prevent disassembly during raising and lowering. Alternatively the sections can be flanged and bolted without slip-fitting. For simplicity, slip-fitting, and any use of flanges and bolts are important details that are ignored in the following analysis. Take the co-ordinate y in the negative vertical direction with origin at the tower top. The horizontal drag per unit height on the tower section at y, $D(y)$, is given by an equation similar to that used in Chap. 4 for aerofoil drag:

$$D(y) = \frac{1}{2}\rho U^2(y) C_d d(y) \tag{10.1}$$

where $d(y)$ is the diameter and C_d is the drag coefficient which is usually specified in the appropriate standard. The variation in U can be determined from formulae similar to those used in Sect. 1.5, e.g. Uys et al. [8]. However, some standards, such as Australian Standard AS1170.2 [12], require the use of the maximum wind speed at all heights. For simplicity, a constant U will be used here. The stress due to the drag is determined by first calculating the shear force V:

$$V = -\int D(y)dy \tag{10.2}$$

and then the moment $M(y)$:

$$M = \int V(y)dy \tag{10.3}$$

with the appropriate boundary conditions at $y = 0$. Equations 10.2 and 10.3 are derived in standard textbooks on structural design. For a linear taper

$$d(y) = d_0 + \frac{d_h - d_0}{h}y = d_0 + d_1 y \tag{10.4}$$

when the deviation due to slip fitting is ignored. d_1 is the taper. Using (10.3) leads to the following expression for the total moment $M(y)$:

$$M(y) = M_0 + T_{\max}y + \frac{1}{2}\rho U^2 C_d\left(\frac{d_0}{2}y^2 + \frac{d_1}{6}y^3\right) \tag{10.5}$$

with the inclusion of $T_{\max}$—the boundary condition on V—and a moment, M_0, acting on the tower. M_0 can arise from a number of causes such as gyroscopic or cyclic loads on the turbine main shaft or a significant overhang of the tower top centre of mass. For simplicity, it is assumed that $T_{\max}$ and M_0 act at $y = 0$. Equation 10.5 is of the form

$$M(y) = a_0 + a_1 y + a_2 y^2 + a_3 y^3 \tag{10.6}$$

which is used in the program to be described. At any height, the maximum bending stress, $\sigma_{b,\max}$, occurs at a distance $d/2$ from the centroid, and is given by

$$\sigma_{b,\max} = \frac{M(y)d(y)}{2I(y)} \tag{10.7}$$

where I, the second moment of area for a regular octagon with wall thickness t, is

$$\begin{aligned} I(y) &= \frac{10 + 6\sqrt{2}}{24(1+\sqrt{2})^3}\left[d^4(y) - (d(y) - 2t)^4\right] \\ &\approx 0.05474\left[d^4(y) - (d(y) - 2t)^4\right] \end{aligned} \tag{10.8}$$

Fig. 10.1 Bending half an octagonal tower section at a factory in China

e.g. Clifton-Smith and Wood [11]. Note that the t is constant within each section—see Fig. 10.1—but can vary between sections. Adding the axial stress due to m_{tt} and the tower self-weight, $m_t(y)$ above y, gives

$$\sigma_{\max} = \sigma_{b,\max} + \frac{[m_{tt} + m_t(y)]g}{A(y)} \tag{10.9}$$

where g is the acceleration due to gravity, and the cross-sectional area, $A(y)$, is

$$A(y) = \frac{2}{1+\sqrt{2}}\left[d^2(y) - (d(y) - 2t)^2\right] \approx 3.3137t[d(y) - t] \tag{10.10}$$

Since the numerator of (10.7) increases as y^4 and the denominator increases as y^3, the maximum stress in each section will occur at the start of its lower slip-fit zone. Furthermore, the increase in the stresses with y usually requires t to be increased in the lower sections. Multi-sectioned towers, therefore, can use a range of section thicknesses to reduce tower mass and cost, but this must be balanced against the requirements of slip fitting, and the availability of steel sheet of appropriate thickness.

Two other important considerations follow from this analysis. The first is that the upwind face of the tower is in tension and the downwind face is in compression, so it is possible for the tower to buckle and this must be designed against. The axial loads in (10.9) will always make the maximum compressive stress larger in magnitude than the maximum tensile stress. Second, the wind direction can vary randomly so it must be assumed that the maximum stress acts on the seam weld between each half-section. This argument applies even for sections of diameter smaller than shown in Fig. 10.1 which can sometimes be fabricated with only one weld. The location of the maximum stress in a weld often reduces the safety factor that can be used.

There are a number of ways to analyse buckling of which two are described here. The first is the ASCE [13] guidelines for regular polygon sections which correlate the results of experiments to determine the limiting values of *a*/*t* where a is the side length. For octagons, $a = d/(1 + \sqrt{2})$. For octagonal sections with specified maximum yield stress F_Y (measured in MPa) the basic condition is:

$$\frac{a}{t}\sqrt{F_Y} < \begin{pmatrix} 680 & \text{for } \sigma_a < 6.9\text{MPa} \\ 630 & \text{otherwise} \end{pmatrix} \tag{10.11}$$

where σ_a is the (axial) stress due to the axial load of the turbine and tower mass. If the inequality holds, then F_Y may be used for design and there will be no buckling. If the left side is larger than the specified limits but still smaller than 960, then the allowable stress considering the local buckling strength of the structure F_a is

$$F_a < \begin{pmatrix} 1.42F_Y\left(1.0 - 4.34 \times 10^{-4}\frac{a}{t}\sqrt{F_Y}\right) & \text{for } \sigma_a < 6.9\text{MPa} \\ 1.45F_Y\left(1.0 - 4.91 \times 10^{-4}\frac{a}{t}\sqrt{F_Y}\right) & \text{otherwise} \end{pmatrix} \tag{10.12}$$

The tests did not extend to values $(a/t)\sqrt{F_y} > 960$. By comparing this correlation to the FEA determination of the linear buckling factor, the ratio of the load required to induce elastic buckling divided by the maximum load, Clifton-Smith and Wood [11] found that (10.12) over-predicted F_a for $t < 4.3 \times 10^{-3}$ m approximately. They suggested the following modification:

$$F_{a,\text{corr}} < \begin{pmatrix} (414t - 0.7842)F_a & \text{for } t < 4.3 \times 10^{-3} \\ F_a & \text{otherwise} \end{pmatrix} \tag{10.13}$$

which will be used here. Many structural codes use a "capacity factor", *CF*, which can be thought of as the inverse of a safety factor. It is the ratio of the maximum calculated stress to the maximum allowable stress, F_A. For present purposes, F_A is the minimum of the yield stress and Eq. 10.12 or 10.13:

$$F_A = \min(F_Y, F_a) \tag{10.14}$$

and

$$CF = \sigma_{\max}/F_A \tag{10.15}$$

where $\sigma_{\max}$ is given by (10.9). Typically, $CF \le 0.6$ if the ultimate tensile strength is used and this value will be used here for illustrative design calculations. Note that the implied safety factor of 1.67 is much less than allowed in the SLM, implying that its use for tower design would result in a massively over-designed tower. This remains a major unresolved issue for small turbine safety assessment.

At least two further quantities must be calculated. The first is the tower top deflection which is obviously the maximum deflection of the turbine and tower. It must remain small to justify the assumption that the structural shape does not alter under the load and hence does not alter the load. The deflection, x in this case, can be calculated according to standard beam theory as,

$$\frac{d^2x}{dy^2} = \frac{M(y)}{EI(y)} \tag{10.16}$$

Anticipating difficulties in analytically determining the double integral, the solution may be sought under the restriction that $t \ll d_0$, in which case

$$I(y) \approx 0.43792t(d_0 + d_1y)^3 \tag{10.17}$$

With this simplification, the double integration of (10.16) using (10.5) and (10.17) may be obtained, for example, using the symbolic mathematics capability of Matlab, but the result appears very cumbersome.

Alternatively, (10.16) can be rewritten as two first order ordinary differential equations and solved using Matlab's built-in Runge-Kutta routines, as was the equation for starting in Chap. 6.

Fig. 10.2 The Aerogenesis 5 kW turbine on its 18 m monopole tower. The gin pole for raising and lowering is attached near the base. The *right side photo* shows the turbine before being raised (photos from Paul Peterson)

Now consider an example design, the octagonal, tapered 18 m galvanized steel tower shown in Fig. 10.2 for an Aerogenesis 5 kW wind turbine. The photo shows the turbine and tower before and after it was raised for the first time with the gin pole still attached. Chapter 12 has a further discussion of raising and lowering. The main tower and turbine parameters are listed in Table 10.2. The base diameter, d_h, and the section thicknesses were chosen iteratively to keep the capacity factors to 0.6 or less. It is important to note that there is no obvious procedure for doing this, so an optimisation method would be valuable, especially if it included a measure of cost in the objective function. The other parameter to note is the turbine thrust, T_{max}, which was determined from the SLM as described in Chap. 9. No load or material safety factor has been applied to this value. The actual tower has extra bracing in the bottom section so that its *CF* for buckling is considerably lower than shown in the Matlab output.

The main parameters for the tower are given in Table 10.2. C_d for an octagonal section is 1.4 from AS 1170.2 [12]. This value is used because it is larger than that in Table 9.5 taken from IEC 61400-2. It is common for different standards to have conflicting requirements and the fact that towers must satisfy IEC 61400-2 as well as relevant local codes means, in practice, that the design will usually be the most conservative possible. This argument should also apply to T_{max} but this would not allow comparison to the FEA values which were based on the data in Table 10.2.

Table 10.1 The main types of towers for small wind turbines

Type	Main advantages	Main disadvantages
Monopole	Aesthetically pleasing	Usually requires more steel More expensive to make and transport
Guyed	Cheap, minimal material Tower natural frequency can be adjusted through guy wire tension Good grounding for lightning	Cannot be used in urban settings Not vandal proof
Sectional lattice	Cheap to transport Assembled on site	Relatively short lifetime as corrosion can start in joints
Tubular lattice	Easy to make, are light and stiff Relatively long lived without galvanising	Less protection for electrical cables

Table 10.2 Parameters for octagonal tapered tower example

Parameter	Value	Parameter	Value	Parameter	Value
h	18 m	N_s	3	U	50 m/s
m_{tt}	170 kg	T_{max}	2200 N	E	200 GPa
d_o	0.17 m	d_h	0.41 m	C_d	1.4
F_y	350 MPa	ρ_{tower}	7800 kg/m^3	m_t	531 kg
t_1, t_2	3.75 mm	t_3	4.3 mm		

Note that the tower mass does not include the slip-fit overlaps, the turbine mounting flange or the baseplate

The listing of the program follows:

```
function tapered_oct_tower(mtt, d0, dh, N, h, Tmax, Umax, M0)
% Determination of maximum capacity factor, tower top deflection
% for an octagonal tower
% mtt - mass of the turbine (tower top) (kg)
% d - diameter across flats (m)
% d0 - minimum diameter (tower top) (m)
% dh - maximum diameter (ground level) (m)
% Fy - yield stress of tower material (Pa)
% t - section thickness (m)
% N - number of sections
% Ns - number of divisions per section
% h - tower height (m)
% Tmax - maximum turbine thrust (N)
% Cd - drag coefficient of octagonal cross-section
% Umax - max. wind speed (m/s)
% M0 - Moment on tower due to turbine
% E - Young's modulus for steel (GPa)
% mt - tower mass (kg)
% rho - density of air (kg/m^3)
% rhot = density of tower material (kg/m^3)

Fy = 350e06; % Material yield stress
Cd = 1.4; E = 200e09 ; rho = 1.2; rhot = 7800; Ns = 10; NsN=Ns*N;
% Coefficients for finding moment as a function of y
d1 = (dh-d0)/h; % d(y) = d0 + d1*y
a1 = Tmax; a20 = 0.5*rho*Umax^2*Cd;
a3= a20*d1/6; a2 = a20*0.5*d0; % Constants for M due to wind & thrust
WindForce= a20*h*(d0+0.5*d1*h);
fprintf('The horizontal force on tower is %7.2f kN\n', WindForce/1000);
Mh = M0 + (a1 + a20*h*(0.5*d0+d1*h/6))*h;
fprintf('The base overturning moment is %7.2f kNm\n', Mh/1000);
t = zeros(N,1);
for i = 1: N; % Find thickness of each section top down ie t
   fprintf(' For section %3i ', i);
   t(i)=0.001*input(' enter thickness in mm;  ');
end

delh = h/NsN; % Divide tower into NsN elements
k=1; y = delh*ones(NsN,1); % y is measured DOWN from top
sigma=zeros(NsN,1);
defl=zeros(N,1) ; mass=zeros(NsN,1); massabove=0.0;
atmax = 0; % Used in ASCE buckling equations
for i = 1: N; %

   for j = 1:Ns; % For each Ns elements per section from top down
      d = d0 + d1*(y(k)-0.5*delh); % Find area at midpoint of section
      mass(k) = rhot*delh*crosssect(d,t(i));
      massabove = massabove + mass(k);
      d = d0 + d1*y(k);
```

```
        faxial = (mtt + massabove)*9.81/crosssect(d, t(i));
        if (faxial > 6.9*10^6)
           fprintf(' Exceeded limit on axial stress; \n')
           return
        end;
        sigma(k)= 0.5*moment(M0,a1,a2,a3,y(k))*d/I(d,t(i))+faxial;
        y(k+1)= y(k)+ delh; k = k + 1;
     end
     check=d/t(i)/(1+sqrt(2)); % For buckling CF
     if (check > atmax); %The maximum occurs at the bottom of section
        atmax =check;
        buckcond=atmax*sqrt(Fy/10^6);
        if (buckcond > 680)
           buckcond = 1.45*Fy/10^6*(1- 491e-06*buckcond);
           if (t(i) < 0.0043) % Correct for small thickness
              buckcond = (414*t(i)-0.7842)*buckcond;
           end
        end
     end
  end
  mt=massabove;
  fprintf('The maximum CF for buckling is is %7.4f\n', max(sigma)/buckcond/10^6);
  fprintf('The maximum CF for stress is %7.4f\n', max(sigma)/Fy);
  fprintf('The tower mass is %7.2f kg\n', mt);
  y(NsN+1)=[];
  y=[0; y]; sigma= [0.5*M0*d0/I(d0,t(1)); sigma];
  % Change y to be UP from base
  % d2x/dy2=M/(EI) is rewritten as two 1st order equations to use
  % Matlab's Runge-Kutta routine ode45 for the deflection, section by section
  IC=[0 0]; % Set initial initial condiitons of no deflection and no slope
  for i = 1:N;
     hspan = [(i-1)*h/N i*h/N]; % Set limits on integration
  % Solve ODE: note how arguments for the derivatives are passed to
  % ode45 and then to defderiv
  [Y, DEFL] = ode45(@(y,x) defderiv(h-y, x,M0, a1, a2, a3, d0, d1, t(i)),hspan,IC);
     IC = [max(DEFL)]; % ICs for next section
  end
  maxdefl=max(DEFL(:,1));
  fprintf('The tower top deflection is %6.3f metres\n', maxdefl);
  kttt = (WindForce+Tmax)/maxdefl; % Find stiffness
  % Natural frequency of tower and turbine
  fprintf('The estimated natural frequency of the tower + turbine is      ...
                                7.3f Hz\n', 0.5/pi*sqrt(kttt/(mt+mtt)));
  fprintf('The estimated natural frequency of the 0.23*tower + turbine...
                              is 7.3f Hz\n', 0.5/pi*sqrt(kttt/(0.23*mt+mtt)));
  end
  function out = I(d, t) % Function for moment of area
  out = 0.43790*t*(d-t)*(d^2-2*d*t+2*t^2);
  end
  function out = crosssect(d, t) % Function for cross sectional area
  out = 3.3137*t*(d-t);
  end
```

```
function out = moment(M0,a1, a2, a3, y) % Function for moment
out = M0+y*(a1+y*(a2+y*a3));
end
function dx = defderiv(y,x,M0, a1, a2, a3, d0, d1, t) % Function for
dx=zeros(2,1);                                        % deflection
dx(1)=x(2);
d = d0 +d1*y;
E = 200e09;
dx(2)=moment(M0,a1, a2, a3, y)/I(d,t)/E;
end
```

Here is a snapshot of the execution of the m-file in the Matlab command window for the IEC SLM Load Case H:

```
>> tapered_oct_tower(170,0.17,0.41,3,18,2200,50,0)
The horizontal force on tower is  10.96 kN
The base overturning moment is  124.65 kNm
For section 1 enter thickness in mm; 3.75
For section 2 enter thickness in mm; 3.75
For section 3 enter thickness in mm; 4.3
The maximum CF for buckling is is  0.6321
The maximum CF for stress is  0.5840
The tower mass is  530.28 kg
The tower top deflection is  0.813 metres
The estimated natural frequency of the tower + turbine is   0.765 Hz
The estimated natural frequency of the 0.23*tower + turbine is   1.185 Hz
>>
```

The comparison of the maximum stress at each height from the program with the FEA results is shown in Fig. 10.3; the simple analysis is reasonably accurate. The maximum axial stress due to the turbine weight and the tower self-weight is 1.2 MPa, which is very small. The important wind loads and tower parameters are either output in the Matlab session copied below the program listing or are consolidated in Table 10.3. The turbine contributes about 20% of the horizontal force and about one-third the base overturning moment, demonstrating why monopole towers need more mass, and are more expensive than the other types.

Table 10.3 demonstrates that the tower top deflection is also reasonably well estimated. The following rows show the natural frequency calculations. Method A approximates the stiffness as the total horizontal force divided by the turbine deflection as calculated above, the natural frequency can be found as the square root of stiffness divided by tower mass. It is easy, however, to show that for a constant diameter tower without a turbine and its thrust, that this very crude method will under-estimate the lowest natural frequency by the factor $1/(0.56\pi) = 0.568$, Exercise 10.14. Method A is more accurate for the combined turbine and tower. This is undoubtedly because the turbine mass is concentrated at the top of the tower and this suggests Method B wherein the effective mass of the tower is set to $0.23m_t$. This factor comes from the standard analysis of the natural frequency of concentrated mass and force on top of a constant diameter column, e.g. pp. 127–128 of Rao [14] which gives the effective column mass quoted above. Furthermore, the axial load due to the turbine weight and tower mass is so small that it will not greatly alter

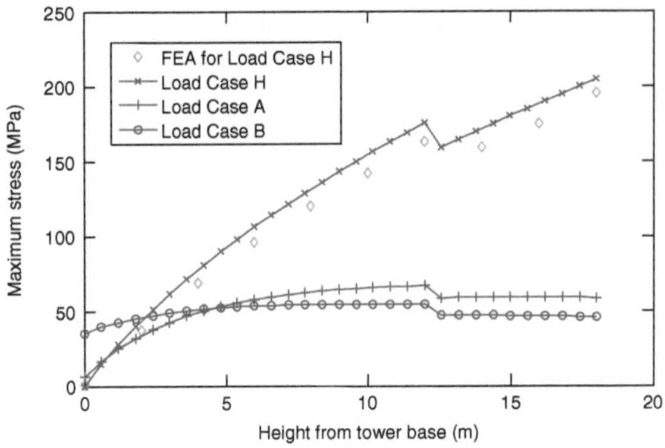

Fig. 10.3 Maximum stress along tower for the IEC SLM load cases calculated by tapered_oct_tower.m and finite element analysis (FEA) results for load case H

the natural frequency. Methods A and B give natural frequencies that lie either side of the FEA value. This value is matched by Method B when the tower contribution is $0.77m_t$, but there is no deeper justification for this value.

Using the ASCE buckling conditions gives a maximum *CF* of 0.623 whereas the linear buckling factor from FEA implies a much lower *CF*. This difference is to be expected since the former is based on experiments on actual octagonal towers.

If the SLM partial safety factors are applied to $T_{\max}$, the maximum *CF*s increase. From Table 9.7, the load safety factor is 3.0, and from Table 9.8, the "fully characterised" material safety factor is 1.1. Their product is 3.3 which is close to twice the inverse of $CF = 0.6$. Therefore the SLM factors can be incorporated into tapered_oct_tower.m by doubling $T_{\max}$ to 4400 N. With this thrust, the maximum *CF* for buckling increases to 0.832 and for stress to 0.768 and the tower would not be safe.

Holmes [3] gives an approximate equation, his (E4), for the lowest natural frequency of a circular tapered tower of varying thickness is

$$n_1 = \frac{\lambda}{2\pi h^2}\left(\frac{EI_h}{\rho_t A_h}\right)^{1/2} \tag{10.18}$$

where

$$\lambda = 1.9\exp(-4d_0/d_h) + 6.65\left[0.9 + (t_0/t_h)^{2/3}\right]^{-1} \tag{10.19}$$

which reduces to the exact formula for untapered towers of constant thickness

$$n_1 = \frac{0.56}{h^{3/2}}\left(\frac{EI}{m_t}\right)^{1/2} \tag{10.20}$$

when $t_0 = t_h$ and $d_0 = d_h$. The comparison with the FE result shows that the approximation is reasonably accurate for this octagonal tower. It is emphasized again, that FEA should be undertaken to accurately determine the natural frequencies and that the approximate methods used here are just that: approximate.

One important use of these calculation is to see whether there is likely to be any resonance caused by blade passing frequencies matching the lowest natural frequency. As for large turbines, see Sect. 7.9.2 of Burton et al. [15], this matching is almost guaranteed for small wind turbines whose blade speed, Ω, is often controlled to match the wind speed to keep the tip speed ratio constant. For example, the two-bladed 5 kW turbine considered here has a rated blade speed of 320 r.p.m. = 5.3 Hz and a blade passing frequency, denoted 1P, of 10.6 Hz. Thus it is almost certain that the turbine will excite the tower natural frequency during some part of its operation. Fortunately, the turbine thrust at a speed corresponding to about 1 Hz is low so that a suitably stiff tower should not show significant resonance and the blades will be stationary when the maximum turbine thrust occurs under Load Case H of IEC 61400-2. Furthermore, wind speed and other variations will usually cause Ω to vary rapidly, so the tower will not have much opportunity to lock-into a perturbation at its lowest natural frequency.

The natural frequency is also required in order to determine whether a static or "dynamic" analysis is required. AS1170.2 [12], for example, allows a static analysis for structures with $n_1 > 1$ Hz but mandates a modified analysis for lower values using increased wind loads. This often requires stiffening the tower.

The discussion and analysis so far has ignored any fatigue loading on the tower, which in the IEC SLM, can only be from Load Case A: Normal Operation. The 5 kW turbine has a design wind speed of 10.5 m/s, and a tip radius of 2.5 m. Using the SLM to determine the magnitude of the variations in shaft thrust gives, with the notation from Chap. 9:

$$\Delta F_{x-\text{shaft}} = \frac{3\lambda_{\text{design}} Q_{\text{design}}}{2R} = \frac{3 \times 7.98 \times 238.73}{2 \times 2.5} = 1143N \tag{10.21}$$

see Eq. 9.7. Following Eq. 9.33, the total number of cycles, n, for a 20 year life, is

$$\begin{aligned} n &= N\Omega_{\text{design}} T_d / 60 \\ &= 2 \times 320 \times 20 \times 365 \times 24 \times 60 \times 60/60 = 6.73 \times 10^9 \end{aligned} \tag{10.22}$$

As mentioned in Chap. 9, Load Case A considers the turbine cycling between 50 and 150% of the design speed. Thus the average thrust is 1143 N and the maximum is 1715 N. The Excel spreadsheet SLM.xls gives the shaft moment against the first turbine bearing as 526 Nm and it will be assumed that this is M_0 acting on the tower throughout the fatigue cycles. The following Matlab session shows the calculation of the stresses resulting from the maximum and mean of the cyclic load for SLM Load Case A (with the thickness and natural frequency lines removed);

```
>> tapered_oct_tower(170,0.17,0.41,3,18,1715,10.5,526)
The horizontal force on tower is   0.48 kN
The base overturning moment is   35.15 kNm
The maximum CF for buckling is is  0.2056
The maximum CF for stress is  0.1899
>> tapered_oct_tower(170,0.17,0.41,3,18,1143,10.5,526)
The horizontal force on tower is   0.48 kN
The base overturning moment is   24.85 kNm
The maximum CF for buckling is is  0.1442
The maximum CF for stress is  0.1332
```

and the results for the maximum of the cyclic load are included in Fig. 10.3. The maximum stress is 70 MPa which is well below the endurance limit for most steels. However, as in Chap. 9, it is necessary to see the effect of the non-zero mean stress which will reduce the fatigue life of structural steel. As in Chap. 9 the effect can be assessed using Gerber's relation and assuming an endurance limit of 200 MPa. Gerber's relation for the allowable maximum stress, σ_n, can be written as

$$\frac{\sigma_n}{200} = 1 - \left(\frac{0.1442 \times 350}{350}\right)^2 = 0.98 \tag{10.23}$$

with the yield stress again used as the ultimate stress. The reduction in the allowable stress is very small. Since 70 MPa $<$ 0.98 $\times$ 200 MPa, the tower will not fatigue.

To complete the design load cases, Fig. 10.3 shows the maximum stress distribution for Load Case B with the gyroscopic moment of 3097 Nm as determined by Eq. 9.11:

```
>> tapered_oct_tower(170,0.17,0.41,3,18,1143,10.5,3097)
The horizontal force on tower is   0.48 kN
The base overturning moment is   27.42 kNm
The maximum CF for buckling is is  0.168
The maximum CF for stress is  0.155
```

Once the overall tower geometry has been set, it is necessary to consider the details such as the access door if there is one, the tower-turbine connection, and base. As would be expected, the base and access door, especially if in the bottom part of the tower, are the most critical. Figure 10.4 shows the detailed FE modeling of the base of the 18 m tower in Fig. 10.2 with the elements indicated along with the von Mises stress distribution. Notice that the high stress levels on the downwind face are mitigated by the baseplate and gussets.

10.3 Optimisation of Monopole Towers

The analysis of the previous section can be extended to determine optimal tower shapes, with the first aim of minimising the tower mass which is closely associated with the cost of manufacture, transport, and foundations. Simple optimisation is only possible for a constant diameter, d, constant thickness, t, tower for which $t \ll d$. With M_0 neglected, (10.5) becomes

$$\sigma'_{\max} = \sigma_{\max} - \rho_t g h \approx \frac{T_{\max} h}{0.87584 t d^2} + \frac{1}{td}\left[\frac{\rho U^2 C_d h^2}{3.50336} + \frac{m_{tt} g}{1.6568}\right] \tag{10.24}$$

on the reasonable supposition that the maximum compressive stress occurs at the base. Any possible problem of matching d to the turbine yaw assembly will be

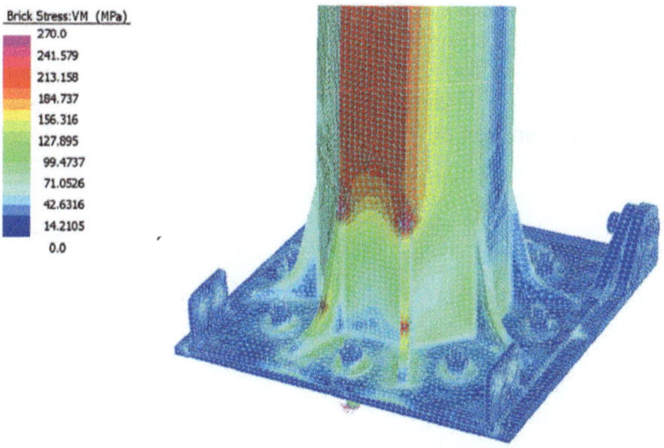

Fig. 10.4 Detailed FEA of the base of the tower in Fig. 10.1. Image provided by Sturt Wilson

ignored. Consider first the artificial case when $T_{\max} = 0$. Then (10.24) is immediately an equation for *td*, which is proportional to the tower mass, and the diameter and thickness can be chosen to avoid buckling. Thus *td* is minimised when the maximum possible *t*/*d* is used to maximise the allowable stress. Including the tower thrust turns (10.24) into a cubic equation for *d* if *d*/*t* is assumed. Fortunately the cubic has only one real root, at least for the single case considered here, and is easily solvable. This is done in the program str_oct_tower.m which is available from the online materials but is not listed. Figure 10.5 shows the variation in mass with *d*/*t* for the same $T_{\max}$, m_{tt}, h, and U as in Table 10.2. Two immediate results are apparent: first, a straight tower is significantly heavier than the non-optimal tapered tower, and second, the minimum mass occurs where $F_{\max}$ deviates from F_y.

The optimisation of tapered towers with different section thicknesses is not possible analytically, but can be tackled by the evolutionary strategy described in Chap. 6 for blade design. A suite of Matlab programs based on the work of Clifton-Smith and Wood [11] is available from the online materials. Previously, Yoshida [16] described the optimization of a large tower using a genetic algorithm. For the present case, the tower genes are d_0, d_h, and the thickness of each of the N_s sections, with example constraints listed in Table 10.4. All important tower information is stored in data structure data created by running tower_opt_setup.m. Then, the main program tower_deopt.m invokes the script t_creator.m to initialise the population. To avoid creating towers with extreme values of *CF*, this script references tapered_oct_tower_opt.m, a straightforward modification to the program listed and discussed earlier in this chapter. Referencing is iterative and continues until all the members of the initial population have a maximum *CF* within reasonable limits. The main modification to tapered_oct_tower.m was to set $CF = 0$ and m_t artificially high whenever $CF_{\max}$ was exceeded, usually in determining the susceptibility to buckling.

The optimisation can combine minimising tower mass, maximising the maximum capacity factor provided it remains below the safe limit, and maximising the tower and turbine natural frequency as estimated by Method B from the last section. The fitness function is modified from Eq. 7.2 to read:

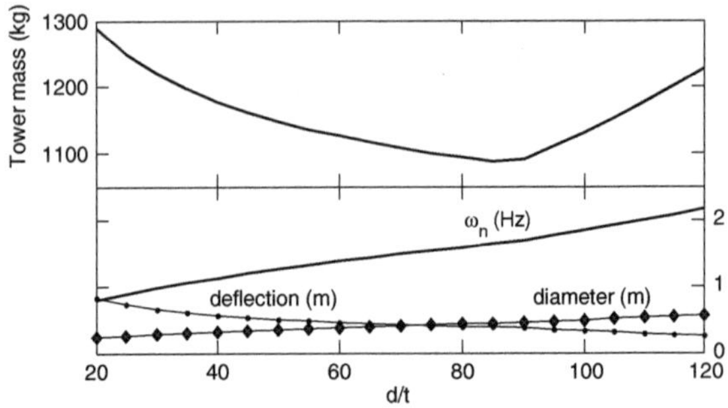

Fig. 10.5 Geometry and parameters of a straight octagonal tower for $h = 18$ m, $U = 50$ m/s, and turbine parameters as in Table 10.2

Table 10.3 Comparison of predicted tower behaviour to finite element results

Parameter	Present value	Value from FEA
$CF_{\text{max stress}}$	0.584	0.557
$CF_{\text{max buckling}}$	0.632	
Linear buckling factor		2.68
Turbine deflection (m)	0.81	0.73
Tower natural frequency—method A (Hz)	1.07	1.51
Tower and turbine natural frequency—method A (Hz)	0.765	0.842
Tower natural frequency Eq. 10.20 (Hz)	1.48	1.51
Tower and turbine natural frequency—method B (Hz)	1.185	0.842

$$\text{fitness}(\boldsymbol{c}_i) = a_m \frac{\min(m_t)}{m_t(\boldsymbol{c}_i)} + a_{CF} \frac{CF(\boldsymbol{c}_i)}{\max(CF)} + (1 - a_m - a_{CF}) \frac{n_1(\boldsymbol{c}_i)}{\max(n_1)} \quad (10.25)$$

For the initial calculations, n_1 was excluded by setting $a_m + a_{CF} = 1$. Note that the maximum *CF* in Eq. 10.25 was 0.60 for all runs, whereas the minimum mass and maximum natural frequency (when used) were determined from each generation in the same manner as for the blade optimisation in Chap. 7. Table 10.4 shows the important parameter and input values for the optimisation; the same *CF* was used for buckling and stress.

Table 10.5 shows that the most critical parameter is the minimum diameter which has a large effect on the optimal mass as well as the distribution of maximum *CF*, Fig. 10.6, largely because reducing $d_{\min}$ increases the maximum *CF* in the top section. In all cases, the maximum *CF* reached the limit of 0.60, and the minimum diameter of the optimal tower converged to $d_{\min}$. This is not surprising given that the top of the tower will contribute significantly to the bending moment at the base, and the magnitude of the moment is determined by the diameter. In practice, $d_{\min}$ is often set by the design of the tower flange and the yaw mechanism which must withstand the gyroscopic moments of Load Case B from Chap. 9, and must allow central clearance for slip rings or the power and other cables to and from the turbine.

The minimum tower mass for $d_0 = 0.17$ m is 45 kg or 8.4% below that for the original tower in Table 10.2, due mainly to the reduced d_h and thickness of the topmost section. It is worth pointing out that restrictions on material supply forced the same thickness steel to be used for the top two sections of the actual 18 m tower. This accounts for about 10 kg of the difference.

Repeated runs for $d_{\min} = 0.10$, and 0.14 produced closely equal results, but Table 10.5 shows some variability in the results for $d_{\min} = 0.17$ due to trade-offs between base diameter and section thickness which did not cause large changes to the mass and none to $CF_{\max}$ or d_0. In practice, as hinted above, there will only be a finite range of available thickness so this variability is unlikely to be significant (Table 10.5).

When using some "local codes" like AS 1170.2 [12], it may be advantageous to include n_1 in the optimisation to decide whether to have a "static" or "dynamic" tower. For example, by setting $a_{n1} = a_{CF} = a_m = 1/3$, and $d_{\min} = 0.17$ m, the evolved m_t was 600 kg, $d_0 = 0.172$ m, $d_h = 0.494$ m, and $t_1, t_2, t_3 = 4.2, 3.4$, and 4.2 mm respectively. Method B gave $n_1 = 1.53$ Hz. Judging from Table 10.2, this should ensure the actual n_1 exceeds the 1 Hz limit in AS 1170.2 [12] for the use of a static analysis with lower wind loads. Whether *increasing* the mass to *lower* the loads is justified is yet another optimisation problem.

Table 10.4 Input parameters for blade design for the optimisation of the 18 m monopole tower. Values not given here are in Table 10.2

Parameter and value	Parameter and value
$Ns = 3$	$CF_{max} = 0.6$
$a_m = 0.5$	$a_{CF} = 0.5$
Maximum $t = 50$ mm	Minimum $t = 2$ mm
Maximum $d = 0.5$ m	Crossover factor, CR = 0.1
Population = 2000	Max. blade age = 20 generations
No. generations = 200	

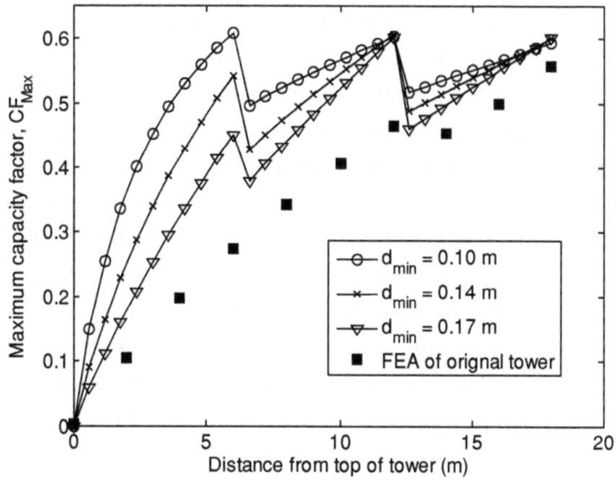

Fig. 10.6 Variation of maximum capacity factor for changes in d_0

Table 10.5 Results of tower optimisation

d_{min} (m)	d_O (m)	d_h (m)	t_1, t_2, t_3 (mm)	m_t (kg)	n_1 (Hz)	Turbine deflection (m)
0.10	0.10	0.373	2.8, 3.6, 4.3	406.9	1.34	0.893
0.14	0.14	0.384	2.5, 3.4, 4.4	439.6	1.28	0.939
0.17	0.17	0.378	2.7, 3.5, 4.8	485.5	1.22	0.966
0.17	0.17	0.384	2.8, 3.6, 4.7	487.0	1.16	0.991
0.17	0.17	0.428	2.6, 3.4, 4.3	497.2	1.12	0.953

10.4 Lattice Towers

Lattice towers are usually triangular or square in planform cross-section. They can be guyed or unguyed but only the latter are considered in this section. Guyed lattice towers often have a constant width and so are similar to guyed pole towers. Both are discussed briefly in the next section. Unguyed lattice towers, like monopoles, often increase in width towards the base to withstand the increasing bending moment. The most famous example is the Eiffel Tower in Paris. Figure 10.7 shows an 18 m high triangular tubular lattice designed for the 5 kW turbine specified in Table 10.2. Table 10.6 gives the main details.

The basic principles of design of lattice towers are similar to those for monopole towers; the highest stresses occur for the extreme wind when the turbine blades are stationary, and buckling is an important and often controlling consideration for the tower components. However, the geometric complexity of lattice towers over monopoles has consequences for the analysis. For example, the wind load is independent of orientation on a monopole but not for a lattice tower, and some standards require an assessment of up to eight different orientations to find the worst case. It may well be that the worst case for one load is in a different orientation than for another. For example, a three-sided lattice tower will usually have the maximum compression at the base of the "back" leg when the wind is normal to the line between the "front" two legs, i.e. in the positive Z direction in Fig. 10.7, but the maximum tension will occur if the wind is in the $-Z$ direction. The lattice itself is sufficiently complex to require a FEA. Simple correlations for the natural frequency are unlikely to be generally valid.

Tubular lattice towers can be fabricated accurately and easily. The 18 m example tower is designed to be manufactured in six 3 m sections using the jig shown in Fig. 10.8, assembled on site, and erected using a separate gin pole with temporary foundations. Table 10.6 shows the thick elements in Fig. 10.8 are made from 60 mm diameter steel water pipe and the thinner bracing is 12.7 mm diameter low Carbon steel. A major reason for the choice of the tubular lattice design rather than a standard lattice tower using bolted angle sections is that the tower is not galvanised and corrosion often starts at the bolted joints.

The wind load on a lattice tower section is usually found by first calculating the solidity of the front face. Solidity is the total projected area of the tower members

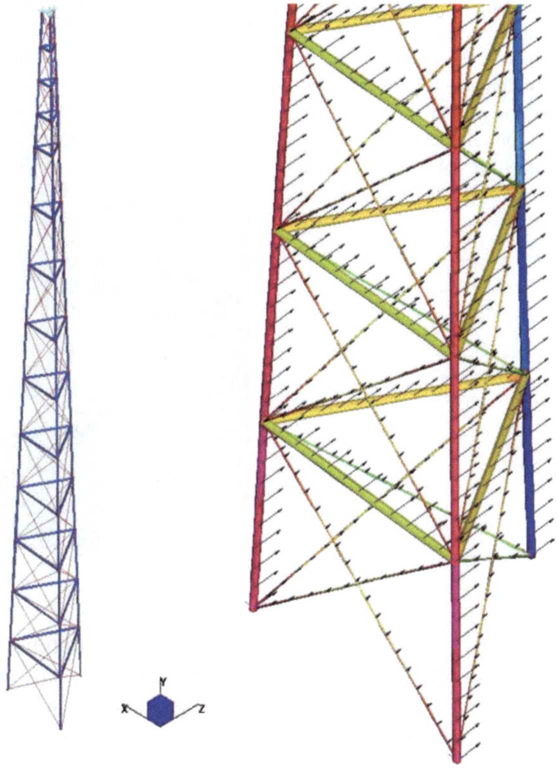

Fig. 10.7 Finite element model of 18 m tubular lattice tower. *At left*, the *blue* (*thick*) *lines* show the horizontal and vertical members and the *red* (*thin*) *lines* the cross-bracing. *At right* is a close up of the bottom section with the relative magnitude of the wind load shown by the *arrows*. The wind is blowing in the positive Z direction. The maximum (compressive) stress of 156 MPa occurs in the lowest "back" leg. Note that the FE model does not include the base plates and connection plates between sections or the gussets. Images from Benn Lakin

Table 10.6 Specifications of lattice tower

General specifications	$h = 18$ m. Tower mass, $m_T = 640$ kg Base distance between legs 1.975 m
Vertical and horizontal members	Low carbon steel. $F_y = 255$ MPa. 60.5 mm diameter pipe. Wall thickness: 3.65 mm
Cross bracing	Low carbon steel. $F_y = 255$ MPa Solid round bar, 12.7 mm diameter

divided by the section area. The drag coefficient then depends on the shape of the members, the direction of the wind, and, for circular members, whether the flow is super- or sub-critical. This distinction relates to whether the boundary layer flow on the member is laminar or mainly turbulent. In some standards, such as AS 3995 [17], the coefficient also depends on whether the lattice is three- or four- legged. From the drag coefficient and extreme wind speed, the load on each section can be determined for each orientation. None of this is difficult, but it is tedious, and can induce the designer to rethink the use of a monopole!

The tower in Fig. 10.7 was designed to use the same materials and have the same base dimensions as towers made by Kijito Windpower for water pumping

Fig. 10.8 Jig to make 3 m long sections of a tubular lattice tower at Kijito Windpower, Kenya. The top section of a tower is in the jig and one of the three base plates for the bottom section is shown by the *white arrow*

wind mills and so no optimisation was attempted. It was fortunate that the maximum FEA stress was 156 MPa which is only slightly lower than that in the monopole tower described in the previous section and corresponds to a *CF* of 0.612. The FEA was done in two parts. First, the overall simple model of the structure shown in Fig. 10.7 was used to find the maximum stresses and the linear buckling factor. Then the base and connection plates with their gussets were modeled in detail, similar to the base of the monopole tower shown in Fig. 10.4.

The remaining task is to assess buckling strength. This was done in the FEA by noting the linear buckling factor, LBF, whose maximum value was 3.53 at the bottom of the back leg in Fig. 10.7. However, it is sensible to check this value against empirical rules available in several forms. Geometric models used in FEA are usually idealised whereas correlations such as those used in Sect. 10.2, include effects of manufacturing defects and inaccuracies. For the circular members of the example tower, either the [13] guidelines or Eurocode 3 [5] can be used. The former use equations similar to those given above for octagonal sections. For axial compression, the limiting stress is

$$F_a = \begin{pmatrix} F_y & \text{for } d/t \le 26203/F_y \\ 0.75F_y + 6550t/d & \text{for } 26203/F_y < d/t \le 82745/F_y \end{pmatrix} \qquad (10.26)$$

For bending:

$$F_b = \begin{cases} F_y & \text{for } d/t \le 41372/F_y \\ 0.7F_y + 12411t/d & \text{for } 41372/F_y < d/t \le 82745F_y \end{cases} \qquad (10.27)$$

where all stresses are in MPa. For axial compression plus bending:

$$\frac{\sigma_a}{F_a} + \frac{\sigma_b}{F_b} \leq 1 \quad (10.28)$$

Equations 10.26 and 10.27 apply to "manufactured" sections rather than those "fabricated from plates…".

To estimate the buckling stress for the back leg of the tower, assume the leg is vertical and supports one-third of the turbine and tower mass. Thus f_a is given by:

$$\sigma_a = 4 \times (170 + 640) \times 9.81 \Big/ \left[3\pi\left(0.0605^2 - (0.0605 - 0.00365)^2\right)\right]$$
$$= 7.87\,\text{MPa} \quad (10.29)$$

Again the axial load due to turbine and tower weight is small. Using $f_b = 156$ MPa from the FEA analysis, and Eq. 10.28 gives

$$\frac{\sigma_a}{F_a} + \frac{\sigma_b}{F_b} = \frac{7.87}{255} + \frac{156}{255} = 0.643 \quad (10.30)$$

so the member is safe from buckling. The Eurocode 3 [5] equations in Annex D of EN 1993-1-6:2007 are more complex. The critical linear meridional (axial in this case) buckling stress is given by

$$\sigma_{Rcr} = 0.605 E C_x t / r \quad (10.31)$$

where r is the mid-radius of the member. The use of (10.31) for studying the buckling of large towers is described in Sect. 7.9.3 of Burton et al. [15]. To fix the value of C_x, it is first necessary to calculate the dimensionless length parameter ω, defined as

$$\omega = l \Big/ \sqrt{(D - t)t/2} \quad (10.32)$$

For the bottom back leg in Fig. 10.7, $l = 1487.5$ mm, $r = 28.425$ mm, and $t = 3.65$ mm, so $\omega = 146$. Since $\omega > 0.5r/t$, the leg is a "long cylinder" and the remaining unknown in (10.31) is

$$C_x = \max\left(0.6, 1 + \frac{0.2}{C_{x,b}}\left[1 - 2\omega\frac{t}{r}\right]\right) \quad (10.33)$$

where C_{xb} depends on the boundary conditions as shown in Table D.1 of Eurocode 3. Taking the ends to be "clamped", that is radially, meridionally, and rotationally restrained, the leg is subject to BC 1 at both ends, and $C_{xb} = 6$. Thus $C_x = 0.6$ from (10.31).

For elastic buckling, the maximum stress from (10.31), must be multiplied by the "imperfection factor" α, which supposedly allows for manufacturing errors and makes these correlations so valuable. It is given by

$$\alpha = \frac{0.62}{1 + 1.91(\Delta\omega_k/t)^{1.44}} \quad \text{where } \Delta\omega_k = \sqrt{rt}/Q \tag{10.34}$$

where Q is the "fabrication quality parameter" which has the value of 40 for Class A or excellent quality, 25 for Class B for high, and 16 for normal quality. Using the last value of Q gives $\alpha = 0.537$. In other words, the inevitable manufacturing defects reduce the maximum allowable stress by nearly one half and this must be considered in any interpretation of linear buckling analysis of the ideal structure. With $\alpha = 0.537$, $\sigma_{Rcr} = 9.32$ GPa from Eq. 10.31. Now according to Eq. 8.17 of Eurocode 3 [5], the "relative slenderness parameter" is

$$\bar{\lambda} = \sqrt{F_y/\sigma_{Rcr}} = \sqrt{255/9320} = 0.165 \tag{10.35}$$

This value is less than the "meridional squash limit slenderness" of 0.20 from Section D.1.2.2 which means that "the characteristic buckling stress" is equal to F_y by Eq. 8.12 of the code rather than σ_{Rcr}. The result, in this case, is the same capacity factor as the ASCE correlations. Both the ASCE and Eurocode 3 [5] correlations imply a linear buckling factor of about half that determined from the FEA, the main difference being that the codes contain some measure of manufacturing imperfections.

Lattice towers can be made of non-circular sections: angles are the most common. They also must be designed to avoid buckling in compression. The relevant equations should be understandable to the reader who has progressed this far, but they are often complex and convoluted and not of prime interest now that the buckling of circular and octagonal members have demonstrated the general design procedure.

The natural frequency of the lattice tower from the FEA analysis is 2.48 Hz which is significantly higher than for the monopole tower. There are a number of correlations for the natural frequency that may be of use. AS 3995 [17] estimates the first natural frequency, n_1 (Hz), as

$$n_1 \approx \frac{1500 w_a}{h^2}\sqrt{\frac{m_1}{m_1 + m_{tt}}} \tag{10.36}$$

where w_a is the average width of the tower, 1.032 m in the present case, and m_1 is the "generalized mass" defined as

$$m_1 = \frac{m_t + m_{tt}}{3}\left[\left(\frac{w_a}{w_h}\right)^2 + 0.15\right] \tag{10.37}$$

where w_h is the width at the base, equal to 1.975 m in this case. Thus $m_1 = 87.43$ kg and $n \approx 2.78$ Hz. Another correlation for lattice towers with ancillaries comes from [18]:

$$n_1 \approx \left(\frac{270}{h}\right)^{2/3}\sqrt{\frac{w_h}{h}} \tag{10.38}$$

which gives $n_1 \approx 2.01$ Hz.

10.5 Guyed Towers

Guyed towers are often the cheapest, Fig. 10.9. Figure 10.10 shows a tubular guyed tower with four guy wires and 5 kW wind turbine being raised using the A-framed gin pole attached to one of the guys. The tower is constructed from sections of constant diameter galvanised pipe and the guy wires are anchored at the

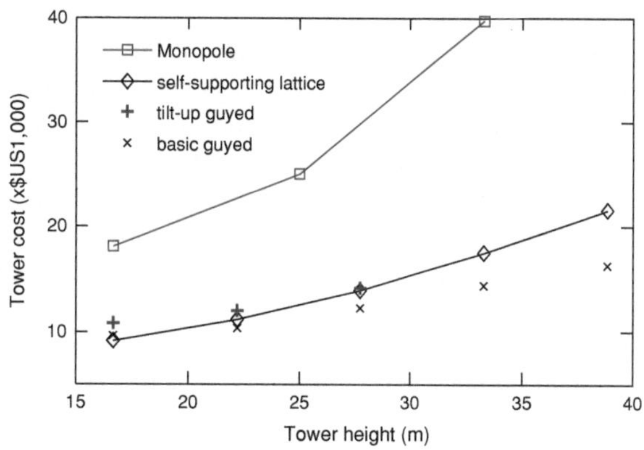

Fig. 10.9 Tower prices for the Bergey 10 kW turbine from www.windpowerunlimited.com accessed April 2010

Fig. 10.10 A 5 kW turbine being raised on a guyed towed tower on French Island, Victoria, Australia

flanges used to join the sections. The turbine is a replacement for one that was dropped during lowering when the guy wires tangled and snapped. The tractor raising the turbine up can be seen on the skyline. At the end of the gin pole is a block and tackle connected to the guy-wire anchor. Guys are often used for raising and lowering a hinged tower. Hinging has a number of advantages, such as lowering tower stress levels by accommodating settling of the foundations, which cannot always be made properly at remote sites. However, it has one major disadvantage: the guy foundations must be designed to withstand the base overturning moment for the worst case of the extreme wind blowing in either direction along the line joining the tractor and the turbine.

With the provision of turnbuckles, it is easy to adjust guy tension for any movement in the foundation and, in principle, to tune for natural frequency variation. Initial guy wire tensions should be around 10% of the allowable stress in the wire [19].

Some useful guidelines for the design of guyed towers are given by Veers et al. [20].[1] They point out that guy-wires must be designed to avoid resonance at blade passing frequencies and showed that the usual guy-wire angle of 35° to the horizontal, which maximises the lateral stiffness, can be relaxed to 45° which reduces the cost and the land area required for the turbine and tower.

Structural analysis of guy-wires is reasonably straightforward, but the tower design usually requires FEA to deal with the concentrated loads imparted by the guys at their attachment pints It can be seen in Fig. 10.10 that attachment in this case occurs at the flange separating the top and middle tower section. This is a common arrangement.

10.5.1 Exercises

1. Derive (10.17) for the approximate moment of inertia when $t \ll d_0$.
2. For a constant diameter octagonal tower with $t \ll d$, and $m_{tt} = 170$ kg, what must the height be for the axial stress, σ_a, the second term on the right of (10.9), to exceed the 250 MPa when $t = 5$ mm and $d = 0.15$ m.
3. Show that the horizontal force, F_{wind}, from a wind load given by Eq. 10.1 on a tapered tower defined by Eq. 10.4 for a constant U is

$$F_{\text{wind}} = \frac{1}{2}\rho U^2 C_d h\left(d_0 + \frac{d_1 h}{2}\right).$$

4. Show that the formula in the previous exercise is dimensionally correct.

[1] This reference is not generally available. The author is indebted to Dr. P. S. Veers for providing copies of the slides used for the conference presentation—there was no formal paper—and for permission to post them with the online materials.

5. Derive the following equation for the base overturning moment, M_h starting with (10.6) for a constant U:

$$M_h = M_0 + T_{\max} h + \frac{1}{4}\rho U^2 C_d\left(d_0 + \frac{d_1 h}{3}\right)h^2.$$

6. Derive F_{wind} and M_h for the case of $U(y)$ as given by Eqs. 1.14 and 1.15.
7. Recast (10.16) for the second derivative of deflection into two first order ODEs and check that the formulation in oct_tower.m is correct. What are the boundary conditions to be used in solving the ODEs? What numerical methods are available to solve these equations?
8. Using Eqs. 10.6 and 10.16, show that the maximum deflection, $\delta_{\max}$, of a monopole tower of constant diameter and wall thickness with no turbine moment (so $a_0 = 0$) is given by

$$\delta_{\max} = \frac{h^3}{EI}\left(\frac{a_1}{3} + \frac{a_2 h}{4}\right).$$

and explain why a_3 does not appear.
9. Use tapered_oct_tower.m to estimate the natural frequency of the example monopole tower without a turbine and so confirm the value of 1.07 Hz in Table 10.2.
10. The 10.66 m tapered tower for the Skystream turbine shown in Fig. 1.2 has a circular cross-section made from 10 gauge steel (3.42 mm). $d_0 = 15.67$ cm and $d_h = 26.56$ cm. Show that the tower mass is nearly 250 kg.
11. The Skystream 2.4 kW turbine has a mass of 77 kg. Estimate the fundamental natural frequency with and without the turbine.
12. Use tapered_oct_tower.m to find the stress distribution, tower top deflection, and natural frequency for the example tower if $T_{\max}$ is replaced by $2T_{\max}$ in accordance with the IEC SLM safety factors.
13. Compare the exact solution from Exercise 10.8 with your results from the program tapered_oct_tower.m. Would you expect the 4th–5th order Runge-Kutta method (ode45 in Matlab) to give the exact answer in this case?
14. For a constant diameter tower with constant thickness, no turbine, but a constant wind loading, use the result of Exercise 10.8 to determine the effective stiffness and show that the Method A from the text leads to the following approximation for the natural frequency

$$n_1 = \frac{1}{\pi h^{3/2}}\left(\frac{EI}{m_t}\right)^{1/2}$$

15. For a constant diameter tower with constant thickness with turbine but no wind loading, determine the effective stiffness and show that the resulting approximation for the natural frequency is

$$n_1 = \frac{1}{2\pi h^{3/2}} \left(\frac{3EI}{m_t + m_{tt}} \right)^{1/2}$$

16. The tower being raised in Fig. 10.10 has a diameter of 30 cm and a wall thickness of 4 mm. It is 13 m high. With the turbine and tower material data as in Table 10.2, calculate the maximum wind speed it could withstand as a monopole.
17. Why is a gin pole (Figs. 10.2 and 10.10) necessary for raising and lowering by cable?
18. Explain why the tower in Fig. 10.10 would exert no overturning moment on its foundations while it is being raised or lowered?
19. Notice that the tail fin on the turbine in Fig. 10.10 is hinged. The hinge has high damping. How might this effect the natural frequency and damping ratio of the turbine yaw response?
20. Towers with a nearly circular cross-section can be subject to peridoic vortex shedding when the wind blows. Tall chimneys, for example, often have spiral additions to break up this shedding which causes a side force that is small but nearly periodic at a frequency f in Hz given by $St = fd/U \approx 0.2$. St is called the Strouhal number. What is the relationship between λ, R, and d if f coincides with 1P? For the two-bladed Aerogenesis 5 kW turbine with a design $\lambda = 9$, and the tower dimensions given in Table 10.2 is f likely to coincide with 1P? What might be the consequences of any coincidence?

21 It is usually argued that interference effects between turbines in close proximity are due to the loss of kinetic energy in the wind received by the downwind turbine(s) after passing through the upwind one(s). How would you estimate the contribution of tower drag to this interference?

References

1. Lavassas I, Nikolaidis G, Zervas P, Efthimiou E, Doudoumis IN (2003) Analysis and design of the prototype of a steel 1-MW wind turbine tower. Eng Struct 25:1097–1106
2. Baniotopoulos CC (2007) Design of wind-sensitive structures. In: Stathopoulos T, Baniotopoulos CC (eds) Wind effects on buildings and design of wind-sensitive structures. Springer, New York
3. Holmes JD (2007) Wind loading of structures, 2nd edn. Taylor & Francis, New York
4. Carril CF, Isyumov N, Brasil RMLRF (2003) Experimental study of the wind forces on rectangular latticed communication towers with antennas. J Wind Eng Ind Aerodyn 91:1007–1022
5. Eurocode 3 (2007) Design of steel structures—Part 1–6: strength and stability of shell structures, En 1993–1–6:2007
6. Geurts C, van Bentum C (2007) Wind loadings on buildings: eurocode and experimental approach. In: Stathopoulos T, Baniotopoulos CC (eds) Wind effects on buildings and design of wind-sensitive structures. Springer, New York
7. SCI (2003) Steel designers manual. In: Davison B, Owens GW (eds) The steel construction institute. Blackwell Science, Oxford

8. Uys PE, Farkas J, Jarmai K, Van Tonder F (2007) Optimisation of a steel tower for a wind turbine structure. Eng Struct 29:1337–1342
9. Kocer FY, Arora JS (1996) Optimal design of steel transmission poles. J Struct Eng 122:1348–1355
10. Dicleli M (1997) Computer-aided optimum design of steel tubular telescopic pole structures. Comput Struct 62:961–973
11. Clifton-Smith M, Wood DH (2010) Optimisation of self-supporting small wind turbine towers. Wind Eng (to appear)
12. AS 1170.2 (2002) Australian standard AS 1170.2:2002 structural design actions—wind actions
13. ASCE (1990) Design of steel transmission pole structures, ASCE manuals and reports on engineering practice no 72
14. Rao SR (2004) Mechanical vibrations, 4th edn. Pearson Prentice Hall, Upper Saddle River
15. Burton T, Sharpe D, Jenkins N, Bossanyi E (2001) Wind energy handbook. Wiley, Chichester
16. Yoshida S (2006) Wind turbine tower optimization method using a genetic algorithm. Wind Eng 30:453–470
17. AS 3995 (1994) Australian Standard AS 3995-1994 design of steel lattice towers and masts
18. Wyatt T (1984) An assessment of the sensitivity of lattice towers to fatigue induced by wind gusts. Eng Struct 6:262–267
19. Gaylord EH Jr, Gaylord CN, Stallmeyer JE (1997) Structural engineering handbook, 4th edn. McGraw-Hill, New York
20. Veers PS, Carne TG, Dohrmann CR (1993) Guy-cable design issues for wind turbine structures, American Society of Mechanical Engineers, Solar Energy Division (publication) SED, vol 14 (available from online materials)

Chapter 11
Generator and Electrical System

11.1 Introduction

Most early small turbines, such as Burne's, mentioned in the Preface, and the famous Jacobs machines from the USA, used DC generators. Modern practice favours three-phase AC permanent magnet generators (PMGs), with induction generators, often just a standard electric motor running "backwards", as a fairly distant second choice. There are novel sources of PMGs, such as domestic washing machines,[1] and an increasing number of them are being made especially for small wind turbines. Many of these come from China which has most of the world's reserves of the rare earth magnetic materials needed for their manufacture.

This change of generators has been driven partly by consumer demand—there are many more AC-powered products than DC-powered—and partly by the advances in power electronics over the last 20 years or so. Nearly all modern generators are three-phase as this increases the ratio of power to weight and produces a more constant shaft torque in comparison to fewer phases. Even for grid connected small turbines, it is usual to rectify the varying frequency and voltage generator power and then invert it to produce AC power of constant voltage and frequency. This has become possible because of the significantly reduced cost and increased capability of modern inverters. Usually this conversion is combined with maximum power point tracking (MPPT) which aims to match the generator power output to the blade characteristics so as to extract the maximum possible power from the wind as its speed varies apparently randomly. MPPT is becoming standard on turbines above the micro category. Many micro-turbines charge batteries and use a much simpler control system. Another more recent innovation is to combine the controller and inverter for grid connection.

This chapter was co-authored by Dr. Peter Freere and Professor Ed Nowicki.

---

[1] For example, http://www.sustainability.ofm.uwa.edu.au/

D. Wood, *Small Wind Turbines*, Green Energy and Technology,
DOI: 10.1007/978-1-84996-175-2_11, © Springer-Verlag London Limited 2011

Filtering is often used to reduce the harmonic distortion of the power supplied to the grid. The inverter must also control the power factor, and oversee the safe shutdown of the turbine if the grid is "lost". In addition the turbine output must be synchronised with the grid.

It is also becoming common for the turbine controller to have a major or exclusive role in over-speed protection. For example, the Skystream 2.4 kW turbine shown in Fig. 1.2, uses shorting of the PMG output instead of aerodynamic controls such as furling and pitching as described in Chap. 8. As with all forms of over-speed protection, there are potential problems with shorting the output. The current or heat generated may demagnetise the magnet and/or burn out the windings.

The need for over-speed control is related to the near-cubic dependence of power on wind speed: wind speeds only slightly higher than the rated speed can cause unsafe levels of blade stress and generator power, and this can happen quickly. Figure 11.1 provides a graphic example of over-speeding from early field tests of the Aerogenesis 5 kW turbine shown in Fig. 10.2 which has a rated wind speed of 10.5 m/s and a disk brake. Before the control system was finalised, the author undertook braking tests with the controller disconnected and the induction generator unexcited. Initially the blades accelerated from rest in a typical high-speed starting sequence. The intention was to apply the brake when the blades reached their rated speed of 320 rpm and document the stopping, which normally takes less than 2 s. However, due to human error (a factor that must not be overlooked when considering safety), the brake was not applied when it should have been at 8 s. By 11 s the blades had reached 730 rpm and the tip speed was over half the speed of sound. It then took 23 s for the brake to stop the blades during which the brake pads were glazed by the heat generated and the brake had to be replaced. During this time, the wind speed did not exceed 150% of the rated speed. While it was gratifying that all other components, including the blades, survived this extreme test, it is not recommended for normal operation. It is likely that the turbine was saved by the rapid drop in wind speed to about 2 m/s at 13 s. At this point the tip speed ratio reached nearly 80, which may be a world record.

This chapter examines all these aspects of the controller and inverter after some further brief comments on PMGs and induction generators. Practical issues of lightning strikes, system protection, and turbine wiring are discussed. Modern controllers may also have an internet connection to broadcast the turbine performance, and allow at least some forms of condition monitoring.

11.2 Generators for Small Turbines

Despite the demise of DC generators, it is worth mentioning them briefly. They are easy to control through the field current and small ones are cheap: even a portable electric drill or vacuum cleaner motor can be made to work as a DC generator. But they have brushes and a commutator, both of which wear out quickly. Since DC

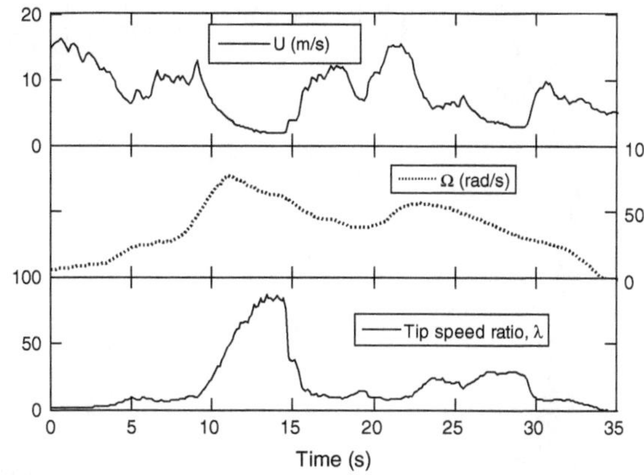

Fig. 11.1 Over-speeding of the Aerogenesis 5 kW wind turbine

generators require a field current to provide the magnetic field, they have more losses than well designed PMGs, and are generally less efficient.

Figure 11.2 shows the very simple structure of a PMG, in this case the Ginlong 500-A that was used with the blade design example in Chap. 7. The modeling of PMGs is well understood, e.g. Gieras and Wing [1] and Proca et al. [2]. The blade holder is normally bolted directly to the input shaft in Fig. 11.1. In addition to being simple, PMGs with many poles—there are 16 poles on the PMG in Fig. 11.2—perform efficiently at low rpm which usually means they do not need a gearbox. The permanent magnets are attached to the rotor which rotates inside the stator carrying the windings. It can be appreciated that detachment of the magnets could cause major problems and be hard to repair. At least some rare earth magnets (e.g. NeFeB) can burn if the protective coating is worn off by striking part of the stator. Some PMG rotors have a lining of fibreglass or iron to keep the magnets in place. All PMGs should have a maximum speed rating below which it is guaranteed to keep the magnets in place for the lifetime of the generator.

Table 11.1 lists the main generator parameters provided by the manufacturer or measured by the methods documented in Kondo [3], and indicates their use in turbine design, modeling, and/or converter design. The manufacturer's specifications for rated power and rpm are lower than those used in Chap. 7 implying that the onus is on the designer to justify the increase in rated conditions.

The maximum efficiency of the generator normally occurs at its highest voltage (which means rated speed, or higher), because for the same power output, the generator current is less. Conduction or "copper" losses increase with an increase in the generator current as the wires get hotter. There are losses from the generator iron being magnetised and demagnetised many times per rotation (called iron losses). These losses depend on the magnetic field strength and the generator speed.

A PMG operated at a constant rpm produces maximum power when the load impedance matches that of the generator. This can be achieved by adding capacitance, but the amount needed is dependent on generator speed (i.e. frequency) and maximum power transfer comes at the price of a dramatic loss

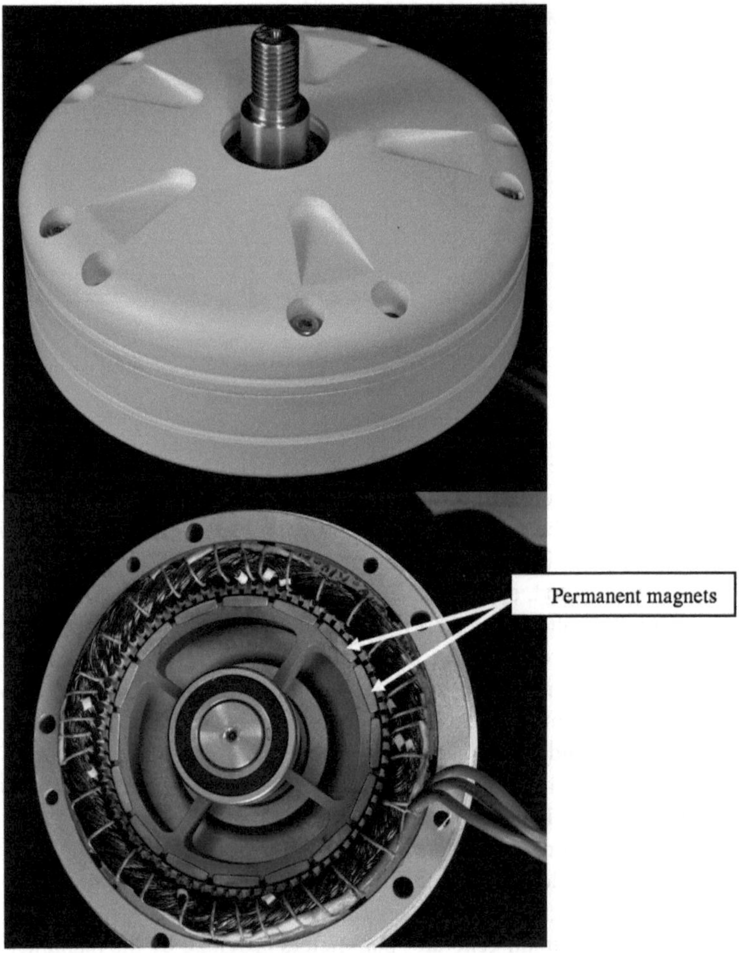

Fig. 11.2 Outside and inside views of the Ginlong 500-A PMG. The casing outer diameter is 250 mm. The maximum shaft diameter is 34 mm. The 16 permanent magnets are attached to the rotor and the back bearing is clearly visible in the bottom photograph. Both photos by Brandon Ferguson

in efficiency. For a wind turbine, which must operate efficiently over a wide range of loads and frequencies, better load matching is needed. For the rectifier/controller–inverter topology shown in Fig. 11.3 matching can be achieved through pulse-width modulation (PWM) as explained in Sect. 11.4. The PMG produces three-phase power of varying voltage and frequency which is normally rectified to DC. If the turbine is charging batteries then they would replace the inverter in the schematic. This possibility may have been in the mind of the designer of the Ginlong 500-A PMG as the DC bus voltage slightly higher than 24 V which is one of the common voltages used in battery banks for remote power systems.

Table 11.1 Main parameters of the Ginlong 500-A PMG

Parameter	Value	Comments
Rated shaft input power	510 W	For blade design
Rated rotor speed	450 rpm	For blade design
Rated input torque	14.8 Nm	For blade design
Cogging torque	0.5 Nm	For blade design
Generator inertia	0.006 kg m^2	Negligible in comparison to blade inertia
Generator weight	12.7 kg	For turbine platform and tower design
Rated stator phase voltage	20 V	rms line to line for rectifier design
Rated stator current	14.4 A	rms line to line for rectifier design
Number of poles	16	Determines operating rpm
Stator resistance	0.35 Ohm	For modeling PMG performance
d-Axis synchronous reactance L_d	3.305 mH	For modeling PMG performance
q-Axis synchronous reactance L_q	3.305 mH	For modeling PMG performance
DC bus voltage at rated speed	25 V	Maximum rectified voltage
DC bus current at rated output	20 A	Maximum rectified current
Permanent magnet flux density λ_{pm}	0.36 V/(rad/s)	For modeling PMG performance

Data from manufacturer and measurements by Mohamed Fahmy and Nacer Benaifa

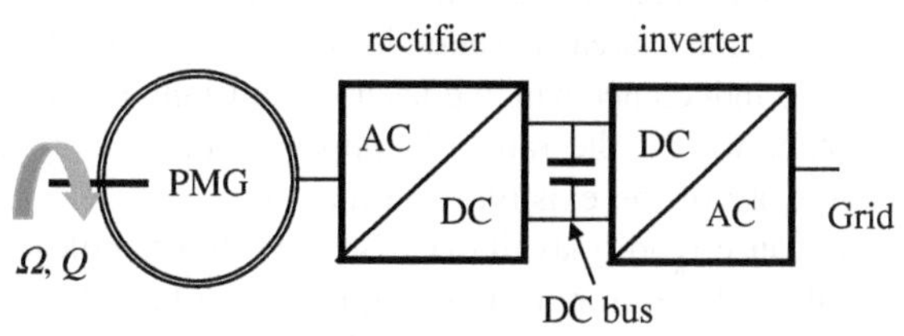

Fig. 11.3 Schematic of a PMG wind turbine feeding the grid. The filtering of the inverter output and possible isolation transformer are not shown

Table 11.1 shows that the cogging torque is small compared to the rated torque even though the ratio is relatively high for this PMG, see Fig. 1.12. The smallness of the cogging torque means that it is only important for starting and low speeds, as considered in Chaps. 6 and 7. Furthermore, it is possible to design the control system to counteract it, e.g. Favre et al. [4], but, of course, that requires the turbine to have started and to be extracting power. Cogging torque is easy to measure, e.g. Zhu [5], and is reasonably well understood, e.g. Guo and Chang [6]. In practice, cogging torque can be reduced significantly, but it is very difficult to eliminate without using iron-less generator designs, e.g. Islam and Sayeed [7].

Some small wind turbines use asynchronous or induction generators (IGs), usually standard induction motors running "backwards". Their main advantages of low cost, no cogging torque, and ruggedness were covered in Sect. 1.8. Because IGs are so common, a description of the basic operating principles can be found in many texts on electrical machines, so only the features important for wind turbine applications are given here. Figure 11.4 shows a typical torque-speed characteristic for an induction machine in terms of the synchronous speed, n_s, at which no torque or power is produced or required. The synchronous speed is the speed at which the rotor is in step with the electrical frequency at the induction motor terminals. The same machine is a motor when operating below n_s and a generator when operating above n_s. The difference is the "slip" which is usually kept to a few percent of the synchronous speed, partly because the slip determines one of the main losses, the so-called I^2R losses dissipated as heat in the rotor. Slip usually increases with decreasing motor size because the rotor resistance is higher for the thinner wire of the cage and hence needs a greater fraction of the rated voltage to induce the required cage current. This is one reason why the efficiencies of small induction motors are usually lower than comparable PMGs, see Table 1.4. There are also I^2R and iron losses in the stator, and smaller amounts of windage and friction losses. Windage refers to the movement of air, either directly by the rotor, or by a fan attached to the generator for cooling.

The synchronous speed is set by the number of poles according to

$$n_s = 120f/N_P \tag{11.1}$$

where the frequency f is either the grid frequency (or the output frequency of a variable frequency drive) for a motor or the rotor frequency for a wind turbine. N_P is the number of poles which is less than 12 for commonly available induction motors. Usually the efficiency decreases with increasing N_P—see Table 1.4—probably because more space is needed for the copper wires, which reduces the cross-sectional area of stator iron and the magnetic coupling. This increases the leakage inductance which does not contribute to the generation of electrical power. In other words, the ratio of copper space to iron space increases which makes poorer use of the existing iron material.

Because readily-available IGs have fewer poles than PMGs, they rotate more rapidly and therefore usually require a gearbox for small wind turbine applications. The word "usually" was carefully chosen because blade frequency increases with decreasing turbine size, so it is possible to design a direct drive wind turbine with an induction generator of say 8–12 poles with a rated power of between 1 and 2 kW. As far as the author knows this has not been done, possibly because these generators are relatively heavy and inefficient.

IGs are only moderately more complex than PMGs. Their rotors have windings rather than magnets, but these windings are shorted and so, like PMGs, do not need brushes or commutators. A variation is a doubly fed induction generator, DFIG,

Fig. 11.4 Torque-speed curve for induction machines

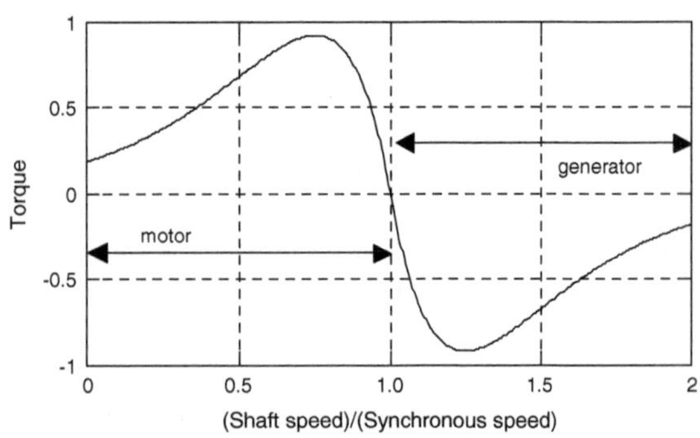

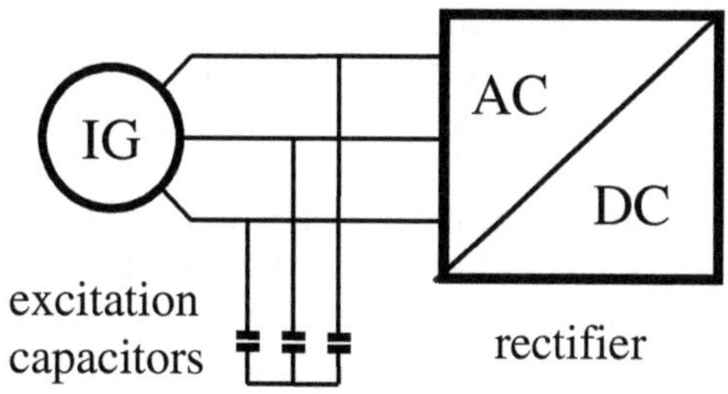

Fig. 11.5 Schematic of an induction generator (IG) with excitation capacitors

which has electrical terminals for the stator *and* the rotor. DFIGs are used on many large turbines. In practice, the main difference from PMGs is that IGs need excitation capacitors to self-excite before they can generate power, e.g. Chan [8], as shown in Fig. 11.5. Otherwise the topology of Fig. 11.2 is the same for both types. The capacitance must be chosen to ensure excitation at a reasonable blade speed by resonance with the induction generator's magnetising inductance; ideally at the end of the starting sequence as discussed in Chap. 6. Unfortunately, the required capacitance can change with cable length and between identically-rated generators from different manufacturers. Furthermore, capacitors are prone to failure, which can result in uneven or no excitation. Then the turbine may overspeed as shown in Fig. 11.1 and it would be foolish to rely on the generator frequency to indicate the blade speed (using a correction for slip). In other words, a zero frequency is no guarantee that the blades connected to an IG are stationary, as it is for a PMG, and a shaft encoder may be required to separately measure Ω. For PMGs, by contrast, the blade rpm is always 60 times the generator frequency in Hz and there is no slip.

11.3 Gearboxes

A gearbox adds complexity, noise, and maintenance. Far more important, however, is that all gearboxes have a resistive torque significantly larger than any resistance in an induction generator when it is not producing power.[2] Gearbox resistive torque, therefore, plays the same role in starting as cogging torque. The analogy extends further: the ratio of gearbox resistive torque on the low speed shaft to rated torque (on the low-speed side) turns out to be comparable to the ratio of cogging to rated torque for PMGs. For example, Fig. 6.10 shows a failed start for the Aerogenesis 5 kW turbine with a gearbox resistance of 1.9 Nm. The rated power occurs at $\Omega = 320$ rpm, giving a rated torque of 149.2 Nm (all values relate to the low-speed or blade side of the gearbox). Thus the ratio of resistive to rated torque is 0.013 which is comparable to those shown in Fig. 1.12 for PMGs.

[2] Since PMG cogging torque would be multiplied by the gearbox ratio it is unlikely that a gearbox would be used in conjunction with a low-pole PMG for small wind turbines.

Gearbox resistance arises partly from oil seal friction on the high speed side, and partly in the bearings for the lay or pinion shaft of an inline gearbox. This is particularly noticeable in gearboxes with pre-loaded taper-roller bearings on that shaft. The small size of the resistive torque makes it unimportant for nearly all gearbox applications and it has not been studied in any detail. In fact, most gearbox manufacturers will scratch their heads if asked about its magnitude.

Gear losses are often divided into speed-dependent and load-dependent components, e.g. Heingartner and Mba [9], of which only the former are of interest for starting. Examples include meshing and windage losses which tend to scale on Ω^2, and generally do not significantly increase the resistive torque.

11.4 Rectifiers, Inverters, and Basic Control

This is the area of greatest and continuing advance in turbines of all sizes. Iov and Blaabjerg [10] briefly describe the development of power electronics for large wind turbines and the history of the various technologies that have increased functionality and programmability while reducing size and cost. They also review the main converter (rectifier and inverter) topologies in use today and provide valuable information on the requirements for grid connection in terms of generator synchronisation, reactive power control, and protection against grid faults. Several excellent books have been written on the electronics and control of large turbines for grid connection, e.g. Munteanu et al. [11] which has a number of case studies involving simulation of a 6 kW turbine as documented in their Appendix A. Many of the basic features of power electronics are independent of turbine size.

One of the simplest power converters is the diode rectifier as diodes turn on and off naturally without the need for any control electronics. A 3-phase diode rectifier is shown in Fig. 11.6, where *A*, *B*, and *C* represent the three phases of the generated AC. The diodes have been numbered 1–6 in the standard manner corresponding to the order in which the diodes turn on. Normally, two diodes are on at the same time; one diode in the top half of the rectifier providing output current, and one diode in the bottom half of the rectifier providing a return path for the DC

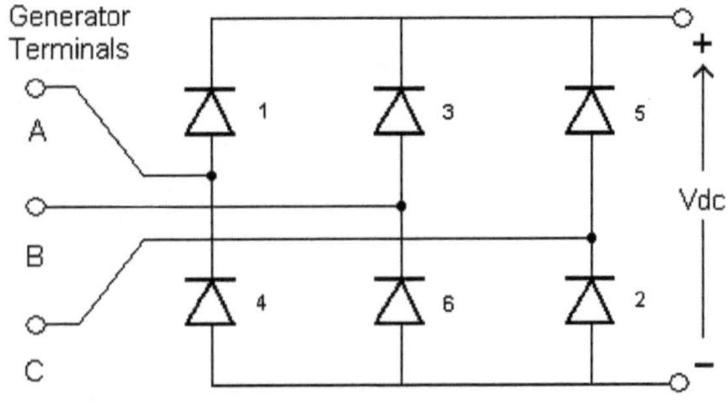

Fig. 11.6 Schematic of a 3-phase diode rectifier

bus current (i.e. if diodes 1 and 2 are on, this is followed by diodes 2 and 3 on, followed by diodes 3 and 4 on, and so forth).

Neglecting the voltage drop across a diode (typically 0.7–0.9 V when the diode is on), and placing a pure resistance across the output (V_{dc} in Fig. 11.6) the average rectifier output voltage is

$$V_{dc} = 3/\pi\left(\sqrt{2}V_{LL}\right) = 1.35V_{LL} \tag{11.2}$$

where V_{LL} is the generator line-to-line rms voltage (for example the rms voltage that would be measured between terminals A and B in Fig. 11.6). Note that V_{dc} is not constant as rectification only makes it non-negative. Equation 11.2 is well known by power converter designers for resistive and inductive type loads. In low power rectifier design, however, the DC bus does not have inductive filtering but rather a simple capacitor filter placed across the DC bus. In this case, the capacitor charges up to a voltage very near the peak of the sinusoidal generator line-to-line voltage, and the DC bus voltage becomes

$$V_{dc} = \sqrt{2}V_{LL} = 1.41V_{LL} \tag{11.3}$$

Whatever the output filtering, a generator will have a wide range of voltages as the wind speed varies, as demonstrated in Fig. 11.7. This has at least two consequences. One is that at low wind speeds, the generator voltage may be so low that the voltage drop across the diodes becomes significant compared to the generator voltage. Rectification will then have a (possibly much) lower efficiency than at rated generator voltage. If the loss of efficiency is too great, the diodes can be replaced with transistor-type switching devices with a much lower voltage drop. An electronic controller, however, is now necessary to turn the transistor switching devices on and off. A second consequence is that at low wind speeds, the DC bus voltage may not be adequate for the inverter to produce the needed AC output voltage. One solution to this is to follow the rectifier with a boost converter.

For small generators in particular, a boost converter is required even for battery charging. From Eq. 11.3 the output voltage of the Ginlong 500-A shown in

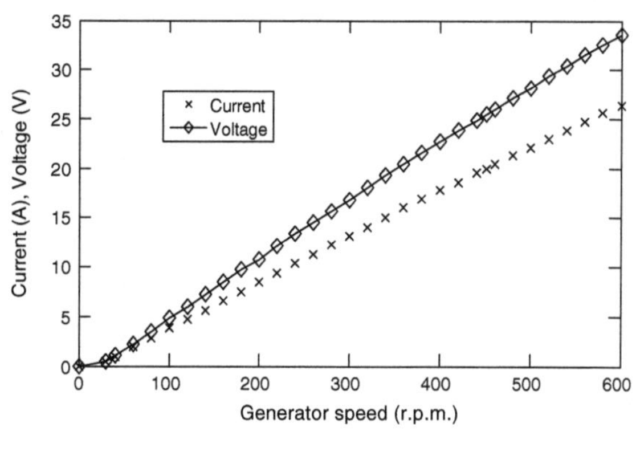

Fig. 11.7 Typical voltage and current output of the Ginlong 500-A PMG

Fig. 11.7 never reaches the 48 V required for most battery systems. As mentioned in the previous section, the key feature of modern power electronics is the ability to use PWM to match generator and load characteristics. In fact PWM is even more powerful as shown by the following example of a boost converter comprising an insulated-gate bipolar transistor (IGBT) or similar device capable of switching very rapidly with little power loss.

Figure 11.8 shows a simple boost converter where the input voltage V_{in} comes from the rectifier and the output voltage V_{out} is the DC bus voltage in Fig. 11.3.

L, *D*, and *C* are the inductance, diode, and capacitor, respectively, and *S* is the IGBT whose switching can be controlled to provide PWM. V_{out} is related to V_{in} by the remarkable equation

$$V_{\text{out}} = V_{dc}/(1 - a) \tag{11.4}$$

which is independent of *L*, *D*, and *C* whose values are fixed by the need to maintain "continuous conduction" and keep within the maximum allowable AC ripple in V_{out}. *a* is the duty cycle, the fraction of time that the switch is shut, e.g. Tafticht et al. [12]. V_{out} is greater than V_{dc} and greater than the peak AC voltage from the inverter if one follows the boost converter. However, if V_{out} is too large with respect to V_{dc} then *a* must be made large and diode *D* must transfer a lot of energy to the capacitor *C* in a short period of time. This stress on the diode and capacitor reduces boost converter efficiency: the maximum practical boost factor $1/(1 - a)$ is about 3. Note on the other hand that if V_{out} is too close to V_{dc} then *a* must be made small with the consequence that switch *S* must quickly transfer significant energy to the inductor *L* which may again reduce boost converter efficiency. A buck-boost converter, having the ability to step-up or step-down the DC bus voltage, may be used in place of the boost converter to overcome switching stresses, [12].

If V_{out} is not fixed, varying *a* can provide MPPT. There are two basic ways of doing this e.g. Munteanu et al. [11]. The first is through some form of "perturb and observe" scheme by which *a* is altered to see if the power output increases. This involves measuring the boost converter output (the DC bus) current and V_{out}. If the power is increased then *a* is varied further in the same direction. If the power decreases, then the variation is reversed. In practice, however, it is often safer to

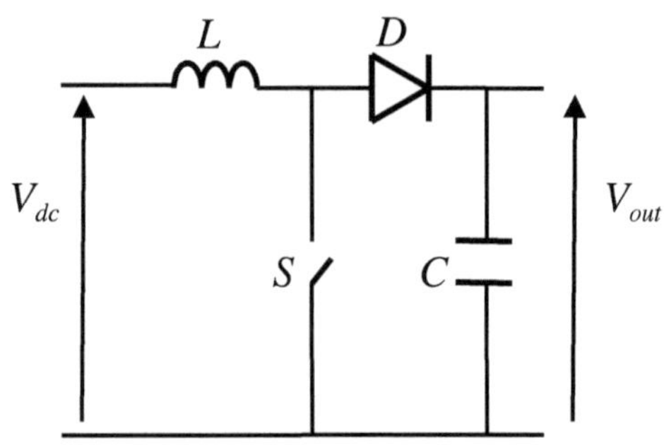

Fig. 11.8 Schematic of a simple IGBT boost converter

have the controller programmed to use a look-up table of optimum power versus frequency, derived from the maximum C_p (for power) and corresponding λ (for frequency) from the performance curve. The power trajectory shown in Fig. 7.1 was constructed in this manner. At any time the actual power and frequency can be compared to the optimum values and the duty cycle altered if there is a significant departure. It is held that this strategy is more robust than the perturb and observe scheme, however it has at least one major shortcoming: the optimal power curve varies with air density so a turbine in a cold climate or on a mountain may need reprogramming. Furthermore, the optimal power curve is derived from the steady performance curve whereas the wind is nearly always unsteady. Given that small turbines have considerably less inertia than large ones—recall Sect. 1.8—there may be considerable benefit in developing unsteady MPPT algorithms. This would require significant improvements in our knowledge of unsteady blade aerodynamics and wake behaviour.

One aspect of MPPT that has yet to be discussed is the rapid reduction in efficiency at part load that is shown in Fig. 7.1 which is typical of small generators, e.g. Di Tommaso et al. [13]. Many small wind turbines operate for considerable time at low generator efficiency, and this is often overlooked in developing MPPT algorithms. Trade-offs may well be profitable in varying the frequency away from the maximum aerodynamic efficiency if there is a correspondingly larger increase in generator efficiency.

An advantage of implementing MPPT in the boost converter rather than the inverter is that battery charging turbines are thereby able to produce more power. However, further power electronics are required to control the battery charging if V_{out} varies and/or the batteries are reaching full charge.

AC output power requires an inverter to convert the DC voltage into a fixed frequency and voltage waveform. A simple single phase "bridge" inverter is shown in Fig. 11.9 where V_{bat} is either the DC voltage from the battery bank or V_{out} from the boost converter in Fig. 11.8. The term "bridge" refers to the output of the inverter which "bridges" the "legs" of the inverter (devices 1 and 3 form one leg while devices 2 and 4 form the second leg of the inverter). No filter is employed, thus the load voltage waveform, V_{load}, is not sinusoidal, Fig. 11.10. Nonetheless, this approach is sometimes taken to make cheap inverters for low

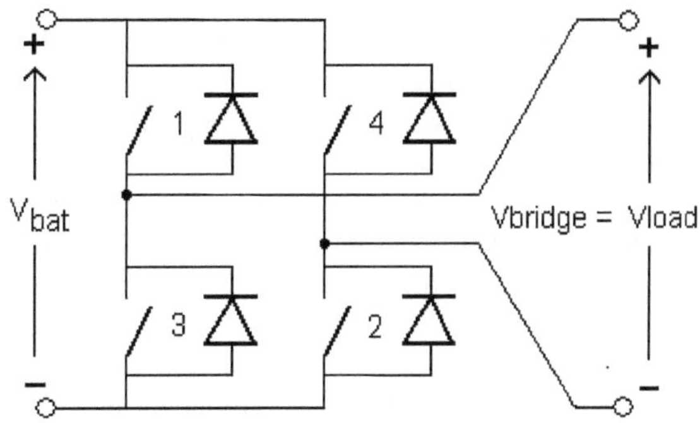

Fig. 11.9 A simple single phase full-bridge inverter

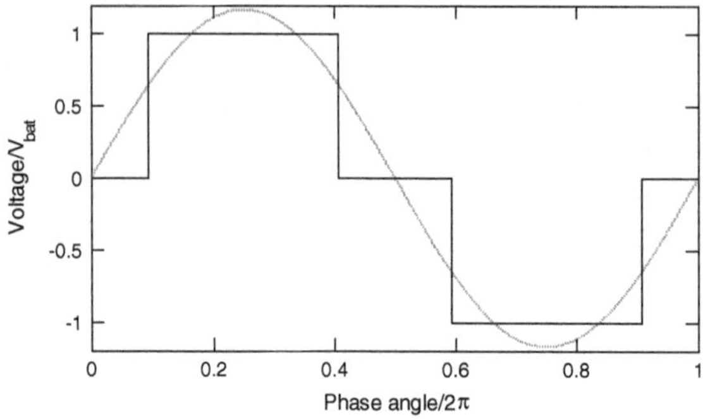

Fig. 11.10 Modified square wave load voltage waveform for a single-phase inverter without a filter. The *vertical axis* is voltage and the *horizontal-axis* is phase relative to the grid frequency

power applications. Most electrical devices, such as computers, home appliances, and motors (with a slight derating), can work well on modified-square-wave-quality AC, but some require very close-to-sinusoidal power. Audio equipment, for example, will often produce a noticeable hum when powered by poor quality AC. To obtain a better approximation to a sinusoidal load voltage waveform a filter could be added and/or a more sophisticated PWM gating algorithm employed to control the timing of the IGBT switching devices. The four switching devices in Fig. 11.9 (or six devices in the case of a 3-phase inverter) can be manufactured individually with or without the feedback diodes; alternatively, the entire set of IGBTs with feedback diodes can be manufactured in a single module.

There is no standard numbering of the switching devices in a single-phase bridge inverter. That used in Fig. 11.9 indicates a simple switch timing to produce the modified square wave load voltage waveform shown in Fig. 11.10. In this figure, the first positive pulse is obtained by gating devices 1 and 2 simultaneously, and the following zero voltage interval is obtained by gating devices 2 and 3. The negative pulse is then obtained by the gating of devices 3 and 4, followed with a zero voltage interval obtained by gating devices 4 and 1. The process is repeated to produce the second positive pulse and so forth.

For the modified square wave of Fig. 11.10, the rms voltage is

$$V_{load,rms} = V_{bat}\sqrt{2\theta_{op}/\pi} \tag{11.5}$$

where θ_{op} may be chosen to minimise, for example, the total harmonic distortion, THD, of the modified square wave (in which case θ_{op} corresponds to 67°). The THD measures the deviation of the actual voltage waveform from the nominally pure sine wave of the grid electricity.

The full-bridge inverter without a filter is very popular for low power inverter systems, However, the application of PWM to the switching devices and adding filtering can make the inverter output nearly sinusoidal. There are two general approaches. For grid-tied inverters, a small inductor may be inserted between the positive inverter output terminal and the grid system voltage (i.e. the inductor is in a series with the grid voltage). The controller is usually designed such that the

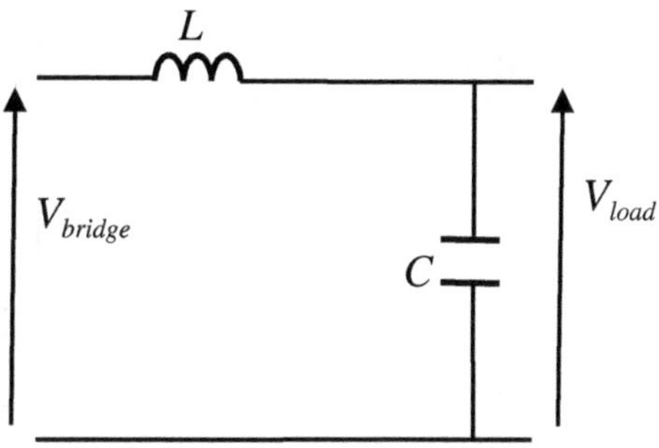

Fig. 11.11 A low-pass LC filter employed to attenuate the switching frequency voltage component of the inverter bridge output resulting in a low distortion near-sinusoidal voltage across the load

square-wave type bridge output voltage is filtered by the output inductor to within a specified rms current for feeding the grid (the rms current value is determined by the available power from the wind turbine or by the system controller in the case of a battery-fed inverter). The THD of the current waveform is typically 5% or less, depending on the appropriate standards for grid connection, for example IEEE 519-1992 for general information and IEEE 1547.2-2008 which specifically covers the interconnection of distributed generators to the grid. The application guide for this standard has valuable background information including "tips, techniques, and rules of thumb".

For off-grid inverter operation, an "LC filter" may be added between the bridge output and the load, as shown in Fig. 11.11. The objective is to produce a near sinusoidal voltage across the load where now the load voltage waveform is typically 5% or less, as determined by consumer product standards such as the German standards association[3] or Underwriters Laboratories Inc in the USA.[4]

Regardless of whether the PWM operated inverter is grid-connected or not, the voltage waveform across the output side of the bridge itself (prior to filtering) is a chopped square wave as illustrated in Fig. 11.12. The width of the pulses is such that the filtered waveform will be a good approximation to a sine wave. Note that the waveform shown in the figure is for illustration purposes only as the switching frequency of the bridge devices is just three times the power frequency (for example 180 Hz for a power frequency of 60 Hz). Practical single-phase PWM inverters have a switching frequency of 15–150 times the power frequency, where the optimal value is a trade off between lower switching losses for the low end and smaller less expensive LC components at the upper end.

An alternative to designing and building an inverter is to purchase one from a number of companies that make inverters especially for small wind turbines. Probably the best known of these are the "Windy Boy" range from SMA.[5]

[3] http://www.din.de/

[4] http://www.ul.com/

[5] http://www.sma.de/en_US/products/wind-power-inverters/

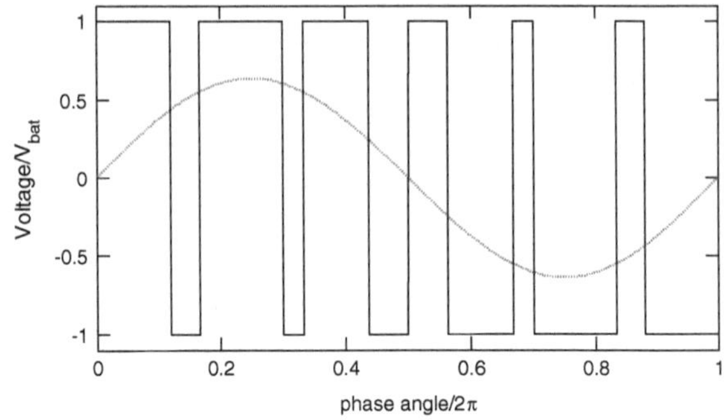

Fig. 11.12 Pulse width modulated (PWM) voltage (*solid line*) and corresponding sine wave (*dashed line*)

11.5 System Protection

Modern microprocessor-based controllers can provide extensive protection for the system and condition monitoring to extend its lifetime. For example, the battery can be protected from being drawn too low and can also be boost-charged, say every month. Some possible faults and appropriate responses are listed in Table 11.2. However, sophisticated protection should be applied intelligently. For example, a turbine should not shut down if smoke is detected from a nearby barbecue.

When constructing or testing a wind turbine, extensive protection is often not desirable as it may affect the turbine behaviour and hence make it difficult to test. Protection against most short term severe damage, such as caused by short circuits, requires the ability to:

1. Disconnect the generator from the rest of the system.
2. Apply and disconnect any dump load.
3. Disconnect the wind turbine from any battery, and to be able to disconnect the battery from any load.
4. Disconnect the inverter from the wind turbine or battery, and to be able to disconnect the inverter from any load.

In all of these cases, the disconnection point also requires over-current protection. Ordinarily, this is done by appropriate circuit breakers, which can be used for both protection and as isolating switches when required, see Fig. 11.13. Note that the figure also shows switches immediately after the generator that can be used to delay power extraction until the blades have reached a suitable rpm after starting. Circuit breakers have differing voltage ratings as to whether they are switching AC or DC currents. Typically, the DC voltage ratings are much lower than the AC ratings.

Basic electrical protection aims to prevent excessive electrical currents. High currents typically cause heating and may melt wires, burn through insulation, and destroy electronic components. In addition, excessive battery-charging current can cause gassing of the batteries and explosions. Excessive current discharging from

Table 11.2 Possible protection issues (response)

Wind turbine	Generator	Battery (if present)	Inverter or load
Over-speeding (slow down or stop)	Over-current (reduce current or disconnect)	Over-charging (stop charging)	Short circuit (disconnect)
Extreme winds (shut down and stop)	Over-voltage (reduce voltage or disconnect)	Excessive voltage (stop charging)	Over voltage (disconnect)
Excessive vibration (reduce turbine speed or stop)	Too hot (reduce current or disconnect)	Too low voltage (disconnect load)	Under voltage (disconnect)
Excessive noise (reduce turbine speed)	Smoke present (disconnect)	Too hot (reduce charging or load current as appropriate)	Too hot (disconnect)
		Smoke present (disconnect)	Smoke present (disconnect and turn off)

the battery can cause over-heating and damage whatever is drawing the current. A lesser effect of the battery over-heating is that its lead plates may expand and bend, thereby shorting. A battery short circuit has been observed on a wind-charging system where it not only started a fire at the short circuit site, but eventually melted the lead battery posts.

When protecting against over-currents, fuses and circuit breakers are commonly used, although it is possible to use controlled power electronics switches or sacrificial components like metal oxide varisters, MOVs. Due to their simplicity and relative reliability, fuses and circuit breakers are often used as the protection of last

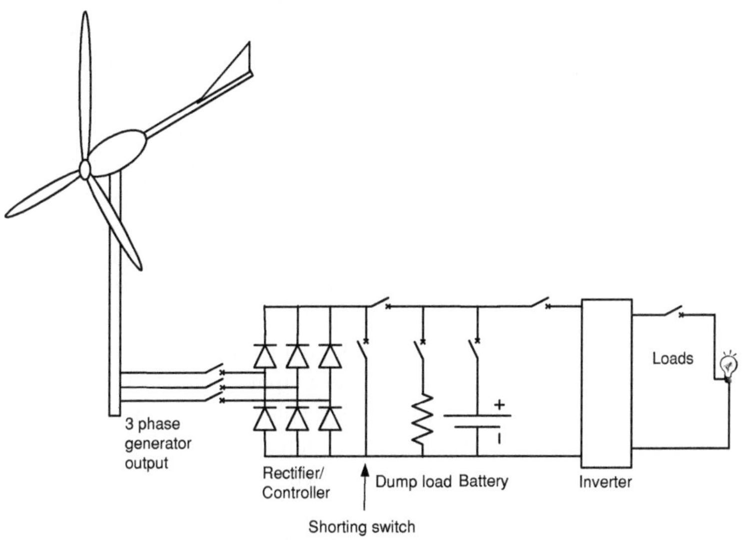

Fig. 11.13 Small turbine system with basic circuit breaker and switch protection

resort. The main advantages of a circuit breaker over a fuse are its ability to be reset, and the ease of turning it off for testing or maintenance purposes.

Fuses and circuit breakers have different specifications when used for AC or DC because of differences in arcing. With AC, any electrical arc will self-extinguish as the current reverses direction. However, DC arcs can establish plasma and hence continue conducting and arcing, potentially with a low voltage drop across the arc and little effect on the current. Any prolonged arc will cause heating, which can melt and ignite the switch, and then spread to the connecting wires.

Table 11.3 shows the tripping characteristics of the ABB pro M compact series of circuit breakers at marginal over-current conditions and severe over-current conditions. Note that these circuit breakers have two ways of releasing—heat and electromagnetic, with different characteristics. The circuit breakers have a nominal tripping current, but also a breaking current capacity. This is the maximum current that the breakers are guaranteed to break (interrupt), typically during a short circuit. The breaking capacity typically ranges from 6000 to 15,000 A.

The tripping characteristics labelled B, C, D, K and Z in Table 11.3 relate to the duration of the currents, the current magnitude, and the length of time it takes for tripping to occur. Of particular interest is the value of the "current hold surges". Surges often occur as equipment is either turned on or connected to the electrical supply from the turbine (e.g. inverters). The circuit breakers should not trip under these conditions, hence higher holding current surge rated circuit breakers may be needed. Class D breakers are commonly used for electric motors.

Fuses are often cheaper then circuit breakers but can be inappropriately replaced by lengths of fencing wire, nails etc. High rupture capacity (HRC) fuses are recommended. Many have very high current breaking capacities of around 80,000 A. All fuses will start to melt at currents above its rating. However, the rupturing of the fuse will take some time, depending on the current and the design of the fuse. This relationship is usually given as an I^2t value, where t is the time to rupture.

The rupture capacity fuse is the maximum current (usually under short circuit conditions) that the fuse is guaranteed to interrupt safely. HRC fuses are usually sand filled to assist with the extinguishing of any arc. Such fuses typically have rupture capacities in the tens of thousands of Amps, even if they are rated at only a few Amps. Hence fuses have the advantage of very high rupturing capacity, low cost, and the ability to maintain shut down until the fault is investigated.

Remote power systems usually have a dump load in the form of a resistive element which may be used for water heating. Normally the dump must be capable of absorbing the continuous maximum power output of the turbine. Despite being crude, dump loads are simple, cheap, and reliable. They can protect the more expensive power electronics especially in systems which are often complex and poorly integrated. Dump loads can also be used for "soft braking"; by dumping part (but not all) of the generated power, the blades can be slowed down for protection or for parking. Few grid connected turbines have separate dump

Table 11.3 Typical circuit breaker tripping details

System pro *M* compact[1]		Technical details: tripping characteristics					MCBs
Tripping characteristics							
Acc. to	Tripping characteristic and rated current	Thermal release[1] current		Tripping time	Electromagnetic release[1] currents		Tripping time
		Conventioel non-trpping c_	Conventional tripping c_		Hold current surges	Trip at least at	
IEC/EN 60898	B 6–63 A	$1.13 \cdot 1_n$		>1 h	$3 \cdot 1_n$		>0.1 s
		$1.13 \cdot 1_n$	$1.45 \cdot 1_n$	<1 h		$5 \cdot 1_n$	<0.1 s
	C 0.5–63 A	$1.13 \cdot 1_n$		>1 h	$5 \cdot 1_n$		>0.1 s
			$1.45 \cdot 1_n$	<1 h		$10 \cdot 1_n$	<0.1 s
	D 0.5–63 A	$1.05 \cdot 1_n$		>1 h	$10 \cdot 1_n$		>0.1 s
			$1.45 \cdot 1_n$	<1 h·		$20 \cdot 1_n$	<0.1 s
DIN VDE 0660/9.82	K 0.5–63 A	$1.05 \cdot 1_n$		>1 h	Not applicable		
			$1.2 \cdot 1_n$	>1 h	$10 \cdot 1_n$		
IEC/EN 60947-2		$1.05 \cdot 1_n$		>2 h		$14 \cdot 1_n$	>0.2 s
DIN VDE 0660			$1.2 \cdot 1_n$	>1 h[2]			<0.2 s
8/69			$1.5 \cdot 1_n$	>2 min[2]			
Part 101			$6.0 \cdot 1_n$	>2 s (T1)			
DIN VDE 0660/9.82	Z 0.5–63 A	$1.05 \cdot 1_n$		>1 h			
			$1.2 \cdot 1_n$	>1 h	Not applicable		
ICE/EN 60947-2		$1.05 \cdot 1_n$		>2 h	$2 \cdot 1_n$		>0.2 s

(continued)

Table 11.3 (continued)

System pro *M* compact[1]		Technical details: tripping characteristics					MCBs
Tripping characteristics							
Acc. to	Tripping characteristic and rated current	Thermal release[1] current		Tripping time	Electromagnetic release[1] currents		Tripping time
		Conventioel non-trpping c_	Conventional tripping c_		Hold current surges	Trip at least at	
DIN VDE 0660			$1.2 \cdot 1_n$	>1 h[3]		$3 \cdot 1_n$	<0.2 s
8/69			$1.5 \cdot 1_n$	>2 min[3]			
Part 101			$6.0 \cdot 1_n$	>2 s (T1)			

Taken from Sect. 10 of http://www05.abb.com/global/scot/scot209.nsf/veritydisplay/47195fae666be225c12576b6003781bb/$File/2CSC400002D0208.pdf (accessed 21 October 2010)

① The indicated tripping values of electromagnetic tripping devices apply to a frequency range of 16 2/3…60 Hz. In the case of diverving frequencies or direct current, see paragraph "Variation of tripping threshold of MCBs according to network frequency" (page 6/7)

② The thermel releases are calibrated to a nominal reference ambient temperature; for Z and K, the value is 20°C, for B and $C = 30°C$. In the case of higher ambient temperatures, the current values fail by ca. 6% for each 10 K temperature rise

③ As from operating temperature (after $1_n > 1$ h or, as applicable, 2 h)

loads but rely on the generator absorbing power and heat when shutting down for "anti-islanding" if the grid connection is lost. Alternatively a smaller dump load can be used.[6] This has the power rating of the turbine but can only absorb the power for a short time while the turbine brakes or is brought to rest by other means.

Example 11.1 The brake on the Aerogenesis 5 kW turbine mentioned in connection with Fig. 11.1 will stop the blades from rated speed in 2 s. What are the power and energy absorption requirements for the dump load for anti-islanding?

Answer The power requirement is obviously 5 kW. A conservative assumption is that 5 kW will be absorbed over the time required to stop so the heat capacity of the dump load is $2 \times 5000 = 10$ kJ.

11.6 Manual Shutdown and Condition Monitoring

Many owners of small turbines rely on the energy produced by the turbine. Hence any method to allow the turbine to generate safely, even with some failures, is desirable. It is therefore important that the controller can be over-ridden in the event of a controller fault. If the controller fails in a dangerous manner (say it leaves the turbine unprotected in strong winds) there should be a way of rescuing the turbine system manually.

For most small turbines, the rectifier must operate in order for the turbine system to function. However, if one phase of the rectifier is damaged, it is possible to isolate the relevant incoming phase, and depending on the type of damage, continue to operate on two phases only. This will cause more vibration and possibly the turbine will need to be partially furled to limit its speed, as its load is now reduced.

If circuit breakers are used, they can be part of a manual override system or for disconnecting parts of the turbine system. The ability to manually initiate soft-braking is also desirable. Some turbines, such as the Skystream, use shorting of the generator output to slow it down. Other generators may either burn out or demagnetise if shorted. However, a shorting switch on the generator output is useful in the absence of a brake, as it allows the wind turbine to remain parked. That is, if the turbine has been slowed by the controller, then it can be stopped by a short circuit and often will not start up in substantial winds. Careful design may be required to prevent accidental parking from near-rated power.

Condition monitoring is a rapidly growing area especially for large off-shore wind turbines. The aim is to simply and cheaply monitor characteristic signals from turbine components in order to assess the machine condition, need for shutdown, and maintenance. Example signals include vibration and generator temperature. Vibration levels can increase significantly if a blade is lost,

[6] The data sheet for one range of these resistors can be found at http://www.welwyn-tt.com/pdf/datasheet/WDBR.PDF (accessed 20 Sept 2010).

the generator loses magnets, or a mounting bolt has come loose. The difficulty, of course, is that the continually varying blade frequency prevents simple analysis of the data. The same is true for gearbox monitoring: for a constant speed gearbox, spectral examination of the vibration levels can often determine which bearing is reaching the end of its life. Advanced analytical techniques are being developed for variable-speed monitoring that will be applied to small turbines in the near future. On the other hand, generator temperature is easy to measure.

Condition monitoring is particularly important for remote turbines, where the maintenance costs can be very high. In conjunction with telemetry-based web-interfaced broadcasting it is possible to acquire and analyse data to schedule specific maintenance for many remote sites. For grid-connected turbines, it is likely that on-line condition and performance monitoring will become very important, especially in those grids that have a high penetration of small-scale renewable generators. The lifetime of many readers of this book could well see general web-based sharing of performance and condition information from small wind turbines and PV installations, combined with resource forecasting, instantaneous demand management, and the like.

11.7 Electrical Wiring

The thickness of any wire used in the electrical system is determined by the heat produced, the need to keep the voltage drop to less than 5%, and the need to maintain adequate system efficiency. These problems are greatest at low voltages, which for small wind turbines, usually occur when battery charging. Table 11.4 gives details of typical low voltage wiring losses and current capacity.

Table 11.4 Copper cable characteristics for typical battery voltages from Appendix C of AS 4509.2 for stand-alone power systems

Current carrying capacity (single phase AC or DC) using HRC fuses or circuit breaker protection						
Conductor size mm^2	Completely surrounded in thermal insulation	Partially surrounded in thermal insulation	Enclosed in air	Unenclosed in air	Buried direct in ground or underground enclosures	Voltage drop in mV/(Am)
1	7	10	12	16	20	44
1.5	10	16	16	20	25	29
2.5	12	20	20	25	32	18
4	16	25	25	62	40	11
6	20	32	32	40	50	7.5
10	25	40	50	50	63	4.5
16	40	63	63	80	100	2.8
25	50	80	80	100	125	1.6
35	57	92	105	115	145	1.3
50	68	110	125	115	170	0.96
70	85	140	155	170	210	0.67

Note that car electrical cables are often only rated at 32 V DC due to their insulation properties. Most cables have a voltage rating, but usually any mains-rated cable will have a voltage rating high enough for most wind turbines. However, the current ratings may be based on intermittent use and should be calculated using Table 11.4.

In addition to resistance, cables have inductance. The inductance reduces the speed at which currents can change, and can cause over-voltage spikes. The inductance can be reduced by keeping the cables near to each other, so as to reduce the enclosed area of the cable.

11.8 Lightning Protection

Lightning strikes tall earthed items, as they provide a lower impedance pathway for the electric charge to earth than the air. Wind turbines can be particularly susceptible to lightning strikes, as the windiest site is often an exposed hilltop. IEC 61400-2 [14] specifically excludes lightning protection as a requirement for small wind turbine blades whereas protection is mandatory for large blades by IEC 61400-24 [15], even though they are made of composites. This standard is unusually comprehensive with details on calculating risk, and advice on protection for large blades with metallic lightning conductors embedded in the blade. Annex N gives valuable information on protection of small grid-connected turbines. Some of the design and operational issues related to lightning protection are discussed by Rachidi et al. [16]. McNiff [17] reports on measurements of strikes on large turbines and gives examples of blade damage.

IEC 61400-24 [15] (Eq. 2) gives an expression for N_d, the annual average number of direct lightning strikes to a turbine as:

$$N_d = N_g A_d C_d \times 10^{-6} \tag{11.6}$$

where N_g is the annual average ground flash density per km^2, A_d is the equivalent "collection area" for lightning strikes on the turbine in m^2, so the factor of 10^{-6} is a conversion factor. The determination of A_d is described in Sect. 7.1 of the standard. Briefly and for the simplest case, $A_d = 9\pi H^2$, where H is the effective height of the turbine, which is the sum of the hub height and blade radius. C_d is an "environmental" factor whose maximum value is 2 for a turbine on a hill.

Values of N_g can usually be obtained from meteorological services.[7] Alternatively, Eq. 1 of IEC 61400-24 [15] gives an estimate for N_g of

$$N_g \approx 0.1 T_d \tag{11.7}$$

[7] For example, the map for Australia can be downloaded from: http://www.bom.gov.au/jsp/ncc/climate_averages/thunder-lightning/ (accessed 30 Sept 2010).

Fig. 11.14 Lightning damage at the trailing edge near the tip of a Westwind 20 kW wind turbine blade at the CSIRO Energy Centre, Newcastle. Photo courtesy of Phil May, Solartec Renewables

where T_d is the number of thunderstorm days per year,[8] another value which can be obtained from meteorological services.

Example 11.2 What is the worst case estimate for the number of lightning strikes on the Aerogenesis 5 kW turbine in Australia?

Answer From the map from footnote 7, the maximum value of N_g is 12 in the tropical north close to Darwin. The tower height is 18 m, and the blade radius is 2.5 m so that $H = 20.5$ m. In the worst case, the turbine is on a hill. From (11.6) the value of N_d is

$$N_d = 12 \times 9\pi \times 20.5^2 \times 2 \times 10^{-6} = 0.285 \text{ direct strikes per year}$$

or nearly six strikes in the 20 year lifetime of the turbine.

Example 11.3 Figure 8 of McNiff [17] gives the maximum number of thunder days per year for the USA as 100 (in Florida). Estimate the number of strikes per year on the turbine of the previous example.

Answer From Eq. 11.7 $N_g \approx 10$, leading to the estimate of nearly five strikes in the 20 year lifetime.

These examples demonstrate a potentially serious risk. Lightning often strikes the blades rather than the nacelle or tower, and can cause major damage even to composite blades as documented in McNiff [17]. Lightning damage to the tip region of a small turbine blade is shown in Fig. 11.14. Most strikes occur near the tip which is the least critical area of the blade's structure and so they can often be repaired easily, provided attention is paid to maintaining the blade mass and centre of mass for balance purposes: the blade in Fig. 11.14 was repaired at the University of Newcastle and was returned to service.

Although small turbine blades do not require lightning protection, it is advisable to include lightning conductors for other components where appropriate: for example, between the generator casing and the tower, if the casing is mounted on

[8] For the USA this information is shown in Fig. 8 of [17].

poorly conducting anti-vibration mounts. National standards on lightning protection in general specify the minimum size, typically 35 mm^2 in cross-section. Power cables and connected equipment should have surge protection such as clamping components, e.g. suitable varistors, transient absorption zener diodes etc. If a turbine is to be installed in an area of high lightning risk, the nosecone and nacelle cover should be metal to protect the generator. The path to ground will then be through the turbine's yaw assembly. The yaw bearing may suffer damage from the lightning current in the form of pitted bearing balls, rollers and races. The bearing may also weld itself in a fixed position.

Generally the tower, regardless of type, provides a sufficient path to ground. Whether earthing is needed at the tower base depends on the foundations and soil type. If, for example, the bolts holding the tower baseplate are electrically connected to the rest of the reinforcing cage, then a concrete foundation can provide sufficient earthing depending on the soil type.

11.8.1 Further Reading

Many references on controllers, rectifiers, inverters and grid connection, as well as a number of important IEEE standards, can be found on the IEE Xplore web site. If your University subscribes to this site you will have free access to all the IEEE-based references cited in this text. Start at: http://ieeexplore.ieee.org/

11.8.2 Exercises

1. The Ginlong 500-A PMG has a maximum efficiency of 73%. For the design example from Chap. 7, how much heat has to be transferred away from the generator at rated conditions?
2. From Figs. 6.2 and 6.5, estimate the rpm at which the generator of the 500 W turbine is switched on.
3. It is now possible to buy micro-inverters intended for standard photovoltaic panels and rated at about 200 W (see for example http://www.enphaseenergy.com/products/). How would you use three micro-inverters for a 600 W wind turbine?
4. If you were designing a 10 kW wind turbine using a PMG, what sensors would you consider installing on the turbine?

References

1. Gieras JF, Wing M (2002) Permanent magnet motor technology. Dekker, New York
2. Proca AB, Keyhani A, EL-Antably A, Lu W, Dai M (2003) Analytical model for permanent magnet motors with surface mounted magnets. IEEE Trans Energy Convers 18:386–391

3. Kondo M (2007) Parameter measurements for permanent magnet synchronous machines. IEEE Trans Electr Electron Eng 2:109–117. doi:10.1002/tee.20117
4. Favre E, Cardoletti L, Jufer M (1993) Permanent-magnet synchronous motors: a comprehensive approach to cogging torque suppression. IEEE Trans Ind Appl 29:1141–1149
5. Zhu ZQ (2009) A simple method for measuring cogging torque in permanent magnet machines, power and energy society general meeting, IEEE. doi:10.1109/PES.2009.5275665
6. Guo Z, Chang L (2008) Calculation and study on cogging torque of small wind turbine PMSG. In: Canadian conference on electrical and computer engineering. doi:10.1109/CCECE.2008.4564603
7. Islam M, Sayeed M (2004) Issues in reducing the cogging torque of mass-produced permanent-magnet brushless DC motor. IEEE Trans Ind Appl 40:813–820
8. Chan TF (1993) Capacitance requirements of self-excited induction generators. IEEE Trans Energy Convers 8:304–311
9. Heingartner P, Mba D (2005) Determining power losses in the helical gear mesh. Gear Technol 22(5):32–37
10. Iov F, Blaabjerg F (2009) Power electronics and machines in wind applications. PEMWA, IEEE. doi:10.1109/PEMWA.2009.5208339
11. Munteanu I, Bractu AI, Cutululis NA, Ceanga E (2008) Optimal control of wind energy systems. Springer, London
12. Tafticht T, Agbossou K, Chériti K (2006) DC bus control of variable speed wind turbine using a buck-boost converter. IEEE Power Eng Soc. doi:10.1109/PES.2006.1709469
13. Di Tommaso A, Miceli R, Galluzzo GR, Trapanese M (2007) Efficiency maximization of permanent magnet synchronous generators coupled to wind turbines power electronics specialists conference. IEEE Power Eng Soc. doi:10.1109/PESC.2007.4342175
14. IEC 61400-2 ed2.0 (2006-03) Wind turbines. Part 2—design requirements for small turbines
15. IEC 61400-24 ed1.0 (2010-06) Wind turbines. Part 24—lightning protection
16. Rachidi F, Rubinstein M, Bermúdez JL, Sola RR, Solà G, Korovkin N (2008) A review of current issues in lightning protection of new-generation wind-turbine blades. IEEE Trans Ind Electron 55:2489–2496
17. McNiff B (2002) Wind turbine lightning protection project. NREL report NREL/SR-500-31115. http://www.nrel.gov/docs/fy02osti/31115.pdf. (accessed 3 Jun 2010)

Chapter 12
Site Assessment and Installation

12.1 Introduction

This chapter considers some of the main aspects of turbine installation: site assessment, optimal tower height, and the loads during erection and lowering which are part of Load Case J of the IEC simple load model. Installation loads are similar to those for transmission structures which are covered by IEEE Std 951-1996. Some brief comments are made on foundations for which a very useful reference is the IEEE Std 691-2001 on transmission structure foundation design and testing.

Site assessment normally involves mapping the wind resource and choosing the best locations for the turbine(s). For large wind farms, anemometer masts up to 50 m high are often used. Monitoring for periods up to 1 year or longer is followed by sophisticated modeling of power production and optimal layout of the individual turbines, e.g. Kusiak and Song [1]. A similar assessment may be prohibitively expensive for a single small turbine so it could be installed without detailed knowledge of the wind resource. If the turbine is the only source of power (or shares that role with PV) then resource assessment may be irrelevant. Nevertheless, neglecting site assessment, or doing it poorly, has often resulted in under-performing turbines, e.g. Encraft [2]. One important aspect of site assessment is selecting the best tower height. Tower costs obviously rise with increasing height as shown by Fig. 10.9 and many manufacturers offer a range of heights, so there will be an optimal height that provides, say, the cheapest electricity per unit capital cost.

The management and cost of installation can often be a major component of a wind turbine project, particularly if there are difficult site features, like reclaimed ground that requires extensive foundations, or a large distance from the turbine to its load. It is not possible to give general guidelines, apart from noting that once the turbine and tower loads—the total mass (vertical load), horizontal force, and base overturning moment—are known, then the design of the tower foundation is no different in principle from any other foundation. (Only lattice towers can possibly cause significant torsional loads on the foundations.) Determination of

D. Wood, *Small Wind Turbines*, Green Energy and Technology,

DOI: 10.1007/978-1-84996-175-2_12,

these loads for extreme winds is covered in Chap. 10, and the corresponding loads for raising and lowering, which can be larger, are considered in this chapter. Wind turbines and monopole towers in particular, tend to generate large horizontal loads in comparison to vertical loads on the foundation. This has lead to some specific foundations, such as screw anchors, e.g. Serato and Victor [3]. An "anchor" is designed to resist pullout rather than compression, so screw anchors are especially appropriate for guyed towers at sites where it is difficult to deliver enough concrete for conventional foundations. IEEE Std 691-2001 also gives information on screw anchors.

12.2 Site Assessment

A common and obvious site requirement for a small wind turbine is that it be sufficiently windy to provide the power required by the load and/or to make a profit for the owner by feeding electricity to the grid. It has been pointed out a number of times that small wind turbines for remote power are usually sited next to, or near, the load they are supplying. Of necessity, a grid-tied small turbine may also be placed close to the owner's house. The only choice left may well be the tower height.

To demonstrate the importance of seeking the windiest possible site, Fig. 12.1 shows the monthly energy output of a Skystream 2.4 kW turbine as a function of average wind speed. Note that the dependence of average power output on average wind speed is closer to linear than the cubic relationship between instantaneous power and wind speed. There are two main reasons for this: first the wind speed is itself distributed over a wide range as demonstrated in Fig. 1.8. Second, the leveling off of the power curve towards the rated wind speed reduces the production at high wind speed, especially for the lower rated speeds of small turbines when compared to large ones. As a rule of thumb, an average hub height wind speed of 5 m/s can be regarded as a good value, so a daily output of 24 kWh is "good" for the Skystream. If the wind speed were reduced to 4 m/s, Fig. 12.1 shows a halving of the output, whereas an increase to 6 m/s increases output by 50%. These changes are

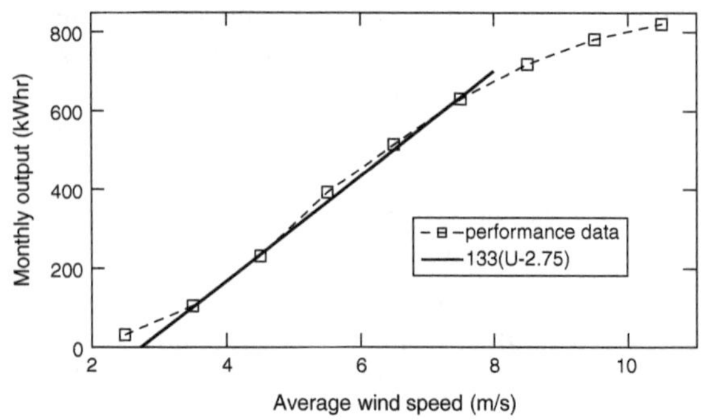

Fig. 12.1 Monthly output for the Skystream 2.4 kW wind turbine. Solid line shows linear approximation as explained in the text

significant and may easily make the difference between an economic and uneconomic project.

Large wind turbines are often sited in "complex terrain" due to hills and valleys and the like. This complicates the resource assessment, partly because the simple equations for the vertical variation of the mean wind speed, Eqs. 1.14 and 1.15 are strictly only valid for fully developed flow over flat terrain. Furthermore, the vertical variation in wind speed usually depends on wind direction, e.g. Ziter and Lubitz [4], and time of day, e.g. Istchenko and Turner [5], because of changes to atmospheric stability. The lower tower heights of small turbines can magnify the effects of terrain especially in the presence of buildings and trees of similar height. Table 12.1 demonstrates some of these features of the wind resource. It lists the daily output from two identical Skystream 2.4 kW wind turbines on 11 m towers over a 16 day period in 2010. The turbines are on the fringe of the city of Calgary (Canada) about 8 km apart. Turbine "A" is on flat ground about 80 m from a large two-storey house. Turbine "B" is about the same distance from a similarly sized house but is on a ridge. Despite these site differences, the average outputs are similar but there are huge variations in daily production. Some of these correlate with wind direction, as output is seriously reduced if the wind flows over the house before reaching the turbine or if the wind flows over the house after passing through the turbine.

It is clear that the wind resource is difficult, and often expensive, to assess, and some compromises are inevitable. There are a number of possible approaches to resource assessment:

Table 12.1 Daily output from identical Skystream 2.4 kW turbines approximately 8 km apart

Date	Daily output from turbine A (kWh)	Daily output from turbine B (kWh)
30 April	0.75	11.15
1 May	1.42	8.48
2 May	8.08	5.16
3 May	17.93	10.93
4 May	0.6	24.56
5 May	6.96	3.39
6 May	4.66	5.65
7 May	11.49	7.32
8 May	38.16	20.8
9 May	21.75	27.79
10 May	3.66	25.28
11 May	22.4	7.13
12 May	7.97	9.95
13 May	13.25	0.98
14 May	28.93	9.29
15 May	13.99	28.65
Average	12.62	12.91

Data from Colin Dumais

- The use of wind atlas data now available for many countries, in the form of average wind speeds at typical heights of 30 or 50 m.[1] Software such as WAsP[2] or 3Tier[3] can interpolate these data to typical turbine hub heights in the presence of complex terrain, trees, and buildings with some limitations due to the "linearity" of the models. WAsP and 3Tier are expensive, so the reader may wish to explore the free alternative Open Wind.[4] Elmore and Gallagher [6] describe an extensive wind resource study for small turbines based on wind atlas data and document some of its pitfalls. Packages such as Meteodyn[5] and Ansys CFX for direct computation of the wind flow are probably too expensive for small turbine siting. A good description of computational studies for micrositing large wind turbines is given by Strack and Riedel [7].
- Rules of thumb or guides from experience. A fascinating on-line example of this is the Danish "wind shading calculator".[6] The U.S. Department of Energy guidelines[7] also give recommendations on minimum distances from the turbine to buildings, trees, and other obstacles. The next section demonstrates that it is possible to undertake only a rudimentary optimisation of turbine height, so the coding of practical experience for choosing height and location will remain a valuable guide for a long time. Furthermore, there has been significant recent effort to improve guidelines using wind tunnel modeling, computational studies, and detailed field measurements, e.g. Brunskill and Lubitz [8].
- The use of software such as Retscreen,[8] Homer,[9] Hybrid2,[10] and SAM[11] for the design of renewable energy systems. All include turbine power production analysis and all are recommended to interested readers. These excellent products are free—at least in their basic form—and all come with extensive wind and solar insolation databases. These databases may be augmented by additional local data from nearby airports, meteorological stations, and pollution monitoring sites. Most of the monthly-averaged wind speeds in these resources were measured at the standard meteorological height of 10 m, so the problem is now to extrapolate to hub height, rather than interpolating wind atlas data. Extrapolation commonly uses the power law (Eq. 1.14) or log law (Eq. 1.15). It is error-prone even for moderate hub heights, e.g. Fig. 1 of Lubitz [9], for the reasons given above.

---

[1] For example, the Canadian wind atlas is at http://www.windatlas.ca/en/ (accessed 7 Aug 2010).

[2] http://www.wasp.dk/ (accessed 7 Aug 2010).

[3] http://www.3tier.com/en/ (accessed 7 Aug 2010).

[4] http://www.awsopenwind.org/ (accessed 25 June 2011).

[5] http://www.meteodyn.com/ (accessed 7 Sep 2010).

[6] http://guidedtour.windpower.org/en/tour/wres/shelter/index.htm (accessed 15 Aug 2010).

[7] http://www.energysavers.gov/your_home/electricity/ (accessed 30 Aug 2010).

[8] http://www.retscreen.net/ (accessed 7 April 2010).

[9] https://www.homerenergy.com/ (accessed 19 May 2010).

[10] http://www.ceere.org/rerl/rerl_hybridpower.html (accessed 2 Jun 2010).

[11] https://www.nrel.gov/analysis/sam/download.html (accessed 7 Oct 2010).

- Some mixture of the above methods. The outcomes of the UK "Warwick trials", Encraft [2], are a good example. This study analysed the performance of a large number of mainly building-mounted turbines and deduced correction factors for use with interpolated wind atlas data.
- Wind speed measurement. It is possible to buy cheap weather stations with cup anemometers and simple data loggers. For reasons that should be apparent by now, the anemometers should be placed as close as possible to the intended hub height, but this is not always easy or cheap. Monitoring for a year at least is recommended, both to document seasonal variations and to allow comparison to nearby long-term averaged data to judge whether the monitoring period is close enough to typical.

Site assessment can include noise and visual impact. There is little to add to the discussion of turbine noise in Sect. 1.3 and the spreadsheet described there and available from the online materials can be used to assess noise at the location(s) of "relevant receiver(s)" to see if the total noise level, i.e. that of the turbine plus "background" significantly exceeds the background level.

Visual impact is a personal issue but can be significant. For example, the turbine in Fig. 10.2 has a cable run of 400 m to the controller in order for the turbine not to interfere with the panoramic views from the farmhouse it powers.

For multiple installations of small turbines, the cost of full site assessment may well be justified. The general procedure is to first identify likely regions of high average wind speeds from wind atlas data or other sources, or monitoring using standard cup anemometers and wind vanes as shown in Fig. 12.2. This mast has an anemometer and wind vane at heights of 9 m and 15 m and is a smaller version of the large masts, often 50 m high, used for large wind turbines. Wind monitoring is usually followed by computer modeling to identify the best locations. A calculated flow over a building is shown graphically in Fig. 12.3.

12.3 Optimum Tower Height

Because many manufacturers offer a range of tower heights, a major part of site assessment is deciding what tower height to use. It is impossible to give detailed recommendations but the following simple analysis shows some aspects that should be considered. Table 12.2 lists the retail prices of the Skystream 2.4 kW turbine and its four tilt-monopole towers as of March 2010.

A fit to this data gives the total turbine price, P_{total}, as

$$P_{\text{total}} = a + bh \tag{12.1}$$

$a =$ \$US6,389 and $b =$ \$US486.8/m. It is reasonable to expect that transport, foundation, and installation costs, which are not included in (12.1), would scale partly on total cost but also have a component specific to each site. These costs are ignored, along with details of project finance which would require estimates of

Fig. 12.2 Anemometer mast for site monitoring for a multiple small turbine installation at Orange, NSW, Australia. Photo supplied by Cyclopic Energy

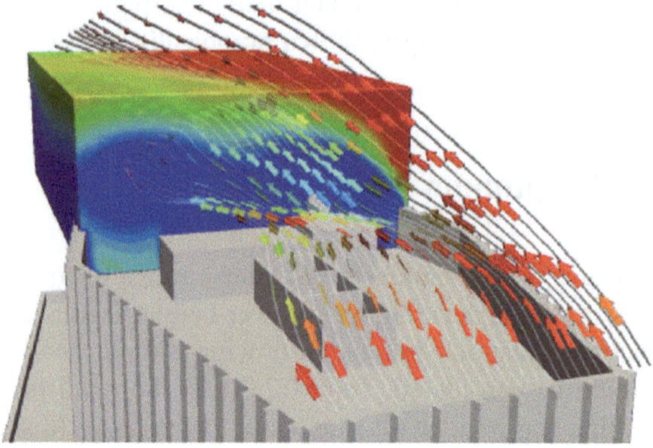

Fig. 12.3 Computer modeling of flow around a building for a proposed small wind turbine installation, image by Cyclopic Energy

Table 12.2 Skystream 2.4 kW turbine and tower prices as of March 2010

Tower height (m)	Retail price ($US)
10.06	11,402
13.72	12,962
16.77	14,245
18.29	15,591

interest rates for borrowed money, and estimates for future electricity prices. Readers interested in these issues can find much useful information in the first-rate user manuals for the Retscreen software mentioned above.

Only the following simple question is considered: what height maximises the average energy output per unit P_{total}? This deceptively simple question can be answered easily only for the very specific conditions considered in Sect. 1.6. If the terrain is flat for a sufficient distance all around the turbine, and the roughness length is constant, then the mean wind speed is given by Eq. 1.14 or 1.15. The former is more useful for our purposes. If the turbine output depends on the average wind speed as shown by the linear fit to the data in Fig. 12.1, with the "offset" wind speed, $U_0 = 2.75$ m/s in this case, then the ratio to be maximised is proportional to

$$\frac{U_{10}(h/10)^m - U_0}{a + bh} \tag{12.2}$$

where U_{10} is the 10 m wind speed, h is in m, and m is the exponent from Eq. 1.14 whose typical values are listed in Table 1.3. Note that the U_0 will be less than the cut-in wind speed, 3.5 m/s for the Skystream, because of the spread of the wind speed distribution. Differentiating (12.2) with respect to h, and equating the result to zero, gives the optimum height, h_{opt}, as the solution to (Fig. 12.4)

$$\frac{ma}{bh_{opt}} + m - 1 + \frac{U_0}{U_{10}}\left(\frac{10}{h_{opt}}\right)^m = 0 \tag{12.3}$$

Increasing m and U_0 and decreasing U_{10} all increase the optimum height. In general, Eq. 12.3 is implicit in h_{opt} and can only be solved analytically for specific values of m, such as $m = ½$ which is of no practical use. This equation was evaluated using Matlab's function fzero in the code shown below

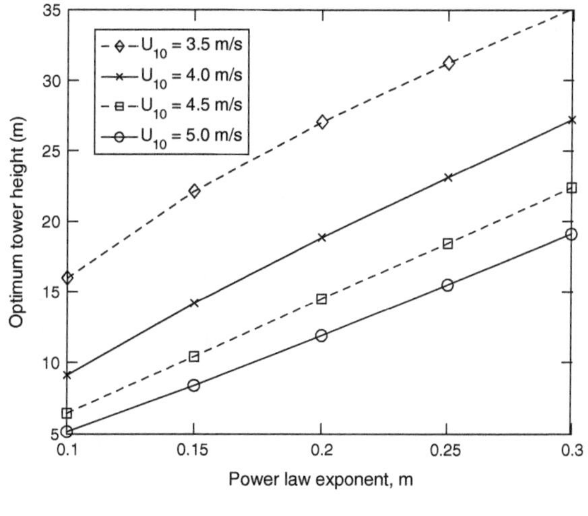

Fig. 12.4 Optimum tower height for Skystream 2.4 kW turbine for varying 10 m wind speeds and power law exponent

```
function out = opt_ht(m,U10, a, b, U0)
out = fzero(@(h) m*a/b/h+ m-1 + U0/U10*(10/h)^m,30);
end
```

where the value “30” in the **fzero** reference is the initial value passed to the root finding algorithm. The results are shown in Fig. 12.4 plotted against m. Although the results are specific to one turbine and its towers, similar calculations for other turbines yield similar results. From Table 1.3, the lowest value of m in Fig. 12.4 is for a very smooth surface, probably not attainable on land, and the largest value corresponds to city centres. For small m the optimum heights are obviously not physical and those for large m and low U_{10} are also questionable because they represent extrapolations of the data in Table 12.2. Nevertheless, it is clear that for rougher sites with low 10 m wind speeds, the optimum tower heights can be very large. It is noted that some manufacturers provide only one tower height of around 10 m.

Typical distributions of Eq. 12.2 with height are plotted in Fig. 12.5 for $U_{10} = 4$ and 5 m/s. In some cases, such as for $m = 0.15$ and $U_{10} = 5$ m/s, there is little change with h, indicating that only a small penalty would be paid for using non-optimum heights. In other cases, such as $m = 0.3$ for the same U_{10}, $h > 20$ m is required for the same conclusion to apply.

So far, the discussion of siting has concentrated on optimising output per unit cost by finding the windiest location and height. This parameter, however, is not necessarily the only one to be optimised and there may be constraints such as visual impact and noise as mentioned above. Windfarm layout is often treated as a multi-dimensional optimisation problem in reducing noise, maximising power production, minimising installation cost etc., e.g. Kusiak and Song [1]. Such studies are not common for small turbine installations, but Professor Ferrer–Marti

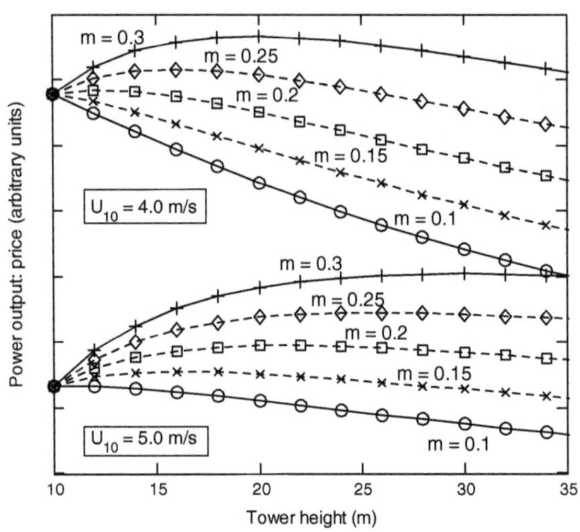

Fig. 12.5 Variation of tower height for Skystream 2.4 kW turbine for varying 10 m wind speeds and power law exponent. The origin for the vertical co-ordinate is zero

and her colleagues[12] have optimised a remote power system in the Peruvian mountains using a number of micro-turbines and demonstrated considerable potential savings in project cost.

Closely related to the estimation of electricity production is life cycle assessment (LCA) which aims to quantify the energy and greenhouse gas cost of producing the electricity. A simple outcome of LCA is the ratio of energy produced over the turbine lifetime to that expended during manufacture, transport, and installation. There are only a few such studies of small wind turbines. Allen et al. [10] show that the ratio varies from 1.7 to 8.8 for a 600 W wind turbine depending on the wind resource. This compares to about 3 for a small PV array, and about 16 for an offshore wind farm. The ratio of CO_2 saved over the turbine lifetime to that expended in its production had a similar range of values.

12.4 Tower Raising and Lowering

It is impossible to give any general information on the transportation and related loads under Load Case J of the IEC SLM, but the major loads during erection can be estimated easily. Figure 12.6 is a simplified diagram of the tower raising or lowering for a tilt-up tower, either guyed or not, using a gin pole. Figures 10.10 and 12.7 show actual raisings of small turbines. For simplicity, it is assumed that the gin pole, of length L, is connected rigidly at the tower hinge point and the cable runs through a pulley close to the end of the gin pole when the tower is vertical. This may also be the anchor point for the separate guy-cable from the end of the gin pole to the tower (which is not shown). θ is the angle of the tower at any height, so that $0 \leq \theta \leq 90°$. β is the angle between the tower and the gin pole, which cannot be significantly less than 90°. The following assumptions are made:

- The raising or lowering is done slowly, so inertial effects are ignored.
- The tower is raised or lowered only during calm weather so there are no wind loads or turbine aerodynamic loads.
- The combined turbine, tower and gin pole has mass m, whose centre lies along the tower at distance l from the hinge point. This distance is not shown in Fig. 12.3.
- The ground is flat.
- No mechanical advantage is used.
- If the tower is guyed, no guy wire is in tension during raising or lowering.
- The tower, gin pole, and cable remain in the same vertical plane so no torsional loads occur, and
- The cable is massless (Fig. 12.6).

[12] http://upcommons.upc.edu/e-prints/browse?rpp=20&order=ASC&value=Ferrer+Mart%C3%AD%2C+Laia&type=author&locale=en.

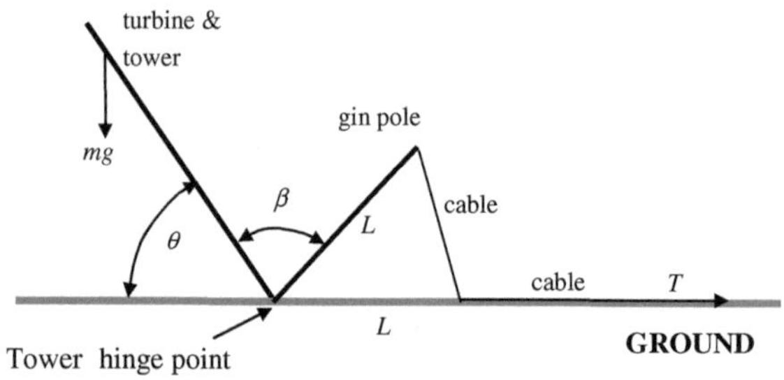

Fig. 12.6 Simplified analysis of cable tensions during raising and lowering a small wind turbine

Fig. 12.7 A Skystream 2.4 kW wind turbine on a monopole tower being raised using a gin pole and winch attached to a four-wheel drive vehicle. Photograph by Rob Falconer, Enmax Corporation

The load of interest is the cable tension T. Once it is known, the stresses in the gin pole, tower, and foundation can be calculated. Taking moments about the tower hinge point and noting that the gin pole, cable, and distance from the pulley to the tower hinge form an isosceles triangle, it is easy to show that the factor f_T relating T to the tower weight, is given by

$$f_T = \frac{TL}{mgl} = \frac{\cos\theta}{\sin((\theta+\beta)/2)} \tag{12.4}$$

and the maximum value of f_T is

$$f_{T,\max} = 1/\sin(\beta/2) \tag{12.5}$$

and so lies between $\sqrt{2} = 1.41$ and about 1.8. The loads required for foundation design are the horizontal force, base overturning moment, and vertical force. On calm days the wind loads are obviously zero. The vertical, compressive load on the foundation during raising and lowering, F_{yb}, is given by

$$F_{yb} = mg\left[1 + \frac{l}{L}\cos\theta\right] \tag{12.6}$$

so that the maximum value, $F_{yb,\max}$, is (Fig. 12.6)

$$F_{yb,\max} = mg(1 + l/L) \tag{12.7}$$

Since l is usually greater than L, the ratio l/L may be termed the "mechanical disadvantage" of the gin pole. The horizontal force on the foundation is

$$F_{xb} = \frac{mgl \cos \theta}{L \tan((\theta + \beta)/2)} \tag{12.8}$$

with a maximum value given by

$$F_{xb,\max} = \frac{mgl}{L \tan(\beta/2)} \tag{12.9}$$

Example 12.1 The 18 m high, 530 kg tower in Fig. 10.2 has its centre of mass 7.46 m from the base. The turbine mass is 170 kg, the gin pole is 6.04 m long, and the angle β as defined in Fig. 12.3 is 80°. Estimate the maximum cable tension in lowering or raising the turbine using Eqs. 12.4–12.9. What are the maximum vertical and horizontal loads on the foundation.

Answer First it is necessary to determine mgl = (530 × 7.46 + 170 × 18) × 9.81 = 68,805 Nm. From Eq. 12.6 $f_{T\max}$ = 1/sin(40°) = 1.56. Therefore $T_{\max}$ = 1.56 × 68,805/6.04 = 17.72 kN which is less than 2 tonnes (= 2 × 1000 × 9.81 kN) in the units commonly used for cable tensions and lifting requirements. From (12.7) $F_{yb,\max}$ = (530 + 170) × 9.81 × (1 + 7.46/6.04) = 15.35 kN which is over twice the vertical force on the foundation when the tower is upright. Equation 12.8 gives $F_{xb,\max}$ = 68,805/(6.04 × tan(40°)) = 13.58 kN which is higher than the maximum horizontal wind force calculated in Chap. 9.

The analysis of raising and lowering shows that minimising the foundation loads, which should minimise foundation costs, is equivalent to minimising the tower mass as was done in Chap. 10.

An important aspect of tower installation and raising for small turbines is related to the fact that most have their centre of mass displaced from the tower axis. This implies that considerable care must be taken to ensure that the tower is vertical once raised, otherwise there could be undesirable gravity effects in the turbine yaw response and furling, if used. Figure 12.8 shows the leveling of a centre-hinged tower for the Aerogenesis 5 kW turbine shown in Fig. 1.2 which was installed using a crane. The foundation is a drilled concrete shaft comprising a 3.5 m long, 900 mm diameter steel reinforcement cage concreted in an auger-dug hole with the square concrete pad visible on the surface. The bolts through the tower baseplate were tied to the cage. In turn, these bolts have nuts below the baseplate which were adjusted as shown to level the tower. Then the gap between the baseplate and the concrete was later filled with grouting.

An alternative steel pile foundation is shown in Fig. 12.9.

Fig. 12.8 Leveling a 5 kW wind turbine tower. This tower is hinged near its mid-height and can be easily swung up and down because the weights visible at the top of the photograph counter the weight of the turbine. Photograph by Paul Peterson

Fig. 12.9 One of a number of small wind turbine pile foundations being driven into the ground. The bevel on top of the foundation seen in the foreground is cut off before attaching the turbine baseplate. Photograph by Rob Falconer, Enmax Corporation

12.4.1 Exercises

1. The noise analysis in Sect. 1.3 considered only a single turbine. How would the Excel spreadsheet presented there be modified to include multiple turbines? Section 10.4.4.2 of Manwell et al. [11] shows how to combine the logarithmic sound power levels from two turbines. This can be easily extended to larger numbers.
2. Redo the analysis leading to Eq. 12.3 using the log law, Eq. 1.15 rather than the power law. Does this alter the conclusions regarding h_{opt}?
3. For $U_{10} = 3.5$, 4, 4.5, and 5 m/s, calculate the average wind speed at $h = 20$ m using the power law for $m = 0.1$, 0.2, and 0.3.
4. Check the derivation of Eqs. 12.4–12.7.
5. Show that Eq. 12.4 can be written as

$$f_T = \sqrt{2}[\cos(\theta/2) - \sin(\theta/2)] = \sqrt{2(1 - \sin\theta)}$$

when $\beta = 90°$. What are the maximum and minimum values of f_T?

6. If the pulley is at distance X from the tower hinge and $\beta = 90°$, show that f_T becomes

$$f_T = \frac{TL}{mgl} = \frac{\sqrt{L^2 + X^2 - 2XL \sin\theta}}{X}$$

How can X be altered to reduce the foundation loads? Is this practicable?

7. If the tower centre of mass was above the tower when it is lying flat on the ground, show that $f_{T,\max}$ is unaltered but that f_T is, in general, reduced.
8. If the mass of the gin pole is included in the analysis, show that $f_{T,\max}$ is increased if $\beta < 90°$.
9. The gin pole in Example 12.1 was made from 100 × 100 × 4 (all mm) steel square hollow section of mass 11.6 kg/m length. Estimate the maximum gin pole contribution to the cable tension and force on foundation during raising or lowering.
10. The ASCE [12] guidelines state that Eqs. 10.11 and 10.12 also apply to rectangular sections in compression. Determine whether the gin pole from Exercise 12.9 will buckle during tower raising or lowering.
11. For the situation in Example 12.1, calculate the maximum pullout forces on the cable pulley.

References

1. Kusiak A, Song Z (2010) Design of wind farm layout for maximum wind energy capture. Renew Energy 35:685–694
2. Encraft (2009) The encraft warwick wind trials project. http://www.warwickwindtrials.org.uk/2.html. (accessed 17 Mar 2010)
3. Serato AB, Victor R (2009) Effects of long-term dynamic loading and fluctuating water table on helical anchor performance for small wind tower foundations. J Perform Construct Facil, ASCE. doi:10.1061/_ASCE_CF.1943-5509.0000013
4. Ziter BG, Lubitz WD (2010) Predicting hub-height wind speed for small wind turbine performance evaluation using tower-mounted cup anemometers. Wind Eng 34:673–699
5. Istchenko R, Turner B (2008) Extrapolation of wind profiles using indirect measures of stability. Wind Eng 32:433–438
6. Elmore AC, Gallagher R (2009) Using regional climate center data to predict small wind turbine performance. Pract Period Hazardous Toxic Radioactive Waste Manage 13:14–19
7. Strack M, Riedel V (2004) State of the art in application of flow models for micrositing. German Wind Energy Conf. http://www.dewi.de/dewi/fileadmin/pdf/publications/Publikations/s07_4_strack.pdf. (accessed 30 Sept 2010)
8. Brunskill AW, Lubitz WD (2010) Development of an empirical obstacle wake model for small wind turbine micrositing. AIAA Paper 2010-828
9. Lubitz WD (2009) Power law extrapolation of wind measurements for predicting wind energy production. Wind Eng 33:259–271

10. Allen SR, Hammond GP, McManus MC (2008) Energy analysis and environmental life cycle assessment of a micro-wind turbine. Proc I Mech E Part A J Power Energy 222:669–684
11. Manwell JF, Mcgowan JG, Rogers AL (2002) Wind energy explained—theory, design and application. Wiley, Chichester
12. ASCE (1990) Design of steel transmission pole structures, ASCE manuals and reports on engineering practice no 72

Index

D. Wood, *Small Wind Turbines*, Green Energy and Technology,
DOI: 10.1007/978-1-84996-175-2, © Springer-Verlag London Limited 2011

MIX
Papier aus verantwortungsvollen Quellen
Paper from responsible sources
FSC® C105338

If you have any concerns about our products,
you can contact us on
ProductSafety@springernature.com

In case Publisher is established outside the EU,
the EU authorized representative is:
Springer Nature Customer Service Center GmbH
Europaplatz 3, 69115 Heidelberg, Germany

Printed by Libri Plureos GmbH
in Hamburg, Germany